Global Morality and Life Science Practices in Asia

Health, Technology and Society

Series Editors: **Andrew Webster**, University of York, UK and **Sally Wyatt**, Royal Netherlands Academy of Arts and Sciences, The Netherlands

Titles include:

Ellen Balka, Eileen Green and Flis Henwood (*editors*)
GENDER, HEALTH AND INFORMATION TECHNOLOGY IN CONTEXT

Courtney Davis (and) John Abraham (*editors*)
UNHEALTHY PHARMACEUTICAL REGULATION
Innovation, Politics and Promissory Science

Gerard de Vries and Klasien Horstman (*editors*)
GENETICS FROM LABORATORY TO SOCIETY
Societal Learning as an Alternative to Regulation

Alex Faulkner
MEDICAL TECHNOLOGY INTO HEALTHCARE AND SOCIETY
A Sociology of Devices, Innovation and Governance

Herbert Gottweis, Brian Salter and Catherine Waldby
THE GLOBAL POLITICS OF HUMAN EMBRYONIC STEM CELL SCIENCE
Regenerative Medicine in Transition

Roma Harris, Nadine Wathen and Sally Wyatt (*editors*)
CONFIGURING HEALTH CONSUMERS
Health Work and the Imperative of Personal Responsibility

jessica Mesman
MEDICAL INNOVATION AND UNCERTAINTY IN NEONATOLOGY

Mike Michael and Marsha Rosengarten
INNOVATION AND BIOMEDICINE
Ethics, Evidence and Expectation in HIV

Nelly Oudshoorn
TELECARE TECHNOLOGIES AND THE TRANSFORMATION OF HEALTHCARE

Margaret Sleeboom-Faulkner
GLOBAL MORALITY AND LIFE SCIENCE PRACTICES IN ASIA
Assemblages of Life

Nadine Wathen, Sally Wyatt and Roma Harris (*editors*)
MEDIATING HEALTH INFORMATION
The Go-Betweens in a Changing Socio-Technical Landscape

Andrew Webster (*editor*)
NEW TECHNOLOGIES IN HEALTH CARE
Challenge, Change and Innovation

Andrew Webster (*editor*)
THE GLOBAL DYNAMICS OF REGENERATIVE MEDICINE
A Social Science Critique

Health, Technology and Society
Series Standing Order ISBN 978–1–403–99131–7   hardback
(*outside North America only*)

You can receive future titles in this series as they are published by placing a standing order. Please contact your bookseller or, in case of difficulty, write to us at the address below with your name and address, the title of the series and the ISBN quoted above.

Customer Services Department, Macmillan Distribution Ltd, Houndmills, Basingstoke, Hampshire RG21 6XS, England

---

Global Morality and Life Science Practices in Asia

Assemblages of Life

Margaret Sleeboom-Faulkner
University of Sussex, UK

First published 2014 by
PALGRAVE MACMILLAN

Palgrave Macmillan in the UK is an imprint of Macmillan Publishers Limited, registered in England, company number 785998, of Houndmills, Basingstoke, Hampshire RG21 6XS.

Palgrave Macmillan in the US is a division of St Martin's Press LLC, 175 Fifth Avenue, New York, NY 10010.

Palgrave Macmillan is the global academic imprint of the above companies and has companies and representatives throughout the world.

Palgrave® and Macmillan® are registered trademarks in the United States, the United Kingdom, Europe and other countries.

ISBN 978–0–230–27483–9

This book is printed on paper suitable for recycling and made from fully managed and sustained forest sources. Logging, pulping and manufacturing processes are expected to conform to the environmental regulations of the country of origin.

A catalogue record for this book is available from the British Library.

A catalog record for this book is available from the Library of Congress.

Typeset by MPS Limited, Chennai, India.

Contents

Preface and Acknowledgements

This is a case study-based textbook for those interested in how different societies deal with developments in the life sciences covering themes ranging from population planning, genetic testing and genomics to human embryonic stem cell research and experimental stem cell therapies. It is especially useful for students interested in science and society, empirical studies of life sciences in Asian societies, and debates in relation to the theoretical conceptualisation of scientific collaboration, biopower, bionetworking, risk, population studies, public deliberation, reproductive choice, eugenics, bioethics and scientific development in countries with differing levels of wealth.

The book contains 17 case studies, nine of which have been co-authored with scholars from China (Suli Sui), South Korea (Seyoung Hwang), India (Prasanna Kumar Patra), Japan (Masae Kato) and the Netherlands (Jyotsna Agnihotri Gupta). Most of the case studies are edited parts of previously published studies, adjusted and reconceptualised for the chapters in this book. The publications on which they draw are referred to in the text.

I hereby would like to express gratitude to *The China Journal*; *The Journal for Health, Culture and Sexuality*; *The Journal for Bioethical Inquiry*; *Biosocieties*; *Social Studies of Science*; *Science as Culture*; *East Asian Science, Technology and Society*; *New Genetics and Society*; *Health, Risk and Society*; *Anthropology and Medicine*; *Medical Anthropology*; and *The Asian Social Science Journal*, in which parts of the case studies have appeared previously.

The case studies have been shortened and rewritten to formulate the arguments about life assemblages made here. Co-authorship was initiated in the context of various research projects, including the Socio-genetic Marginalisation in Asia Programme (SMAP, International Institute for Asian Studies, Leiden–University of Sussex) and International Science and Ethics Collaborations (ISBC, a joint research project of the Universities of Cambridge, Durham and Sussex), with added input from the Bionetworking in Asia projects (Sussex University, Brighton, and Peking Union Medical College, Beijing). I worked with the co-authors on the original publications, and the co-authors have kindly checked and updated the case studies for the book.

The book has benefitted from the financial, administrative and social support of many institutions and people. There is insufficient room to list them all here. Funding was provided by the Netherlands Organization for Scientific Research (NWO), the International Institute for Asian Studies (IIAS), the University of Leiden, the Economic and Social Research Council (ESRC: RES-350-27-0002; RES-062-23–0215; RES-062-23-2990), and the European Research Council (ERC: 283219).

I would like to thank Alex Faulkner for his support and comments on an early version of Chapter 1. André Celtel and Lee Bowers kindly read the text. The book makes use of concepts and references from Hindi, Chinese *Putonghua*, Japanese and Korean sources. The transcription system used for Chinese is pinyin, unless the concepts are better known using alternative spellings. I would also like to thank the series editors for their encouragement, and the reviewers for their helpful comments on an earlier version.

I would like to explain my use of the words 'bioethics' and 'bioethical' here to prevent misunderstanding. Unless otherwise indicated, I do not use these concepts normatively, but attributively, as in 'bioethics institutions' and 'bioethical regulation.' I use the attributive noun *bioethics* before 'committee', 'institution', 'discussion', 'controversy', 'group', 'debate', 'discourse', 'regime', 'campaign', 'expert' and 'policy', cases in which the use of the adjective *bioethical* could be interpreted as being predicative. I use the attributive adjective *bioethical* before 'governance', 'procedures', 'capacity building' and so forth, as the meanings of these more commonly used concepts are widely known. Finally, I use *bioethical* rather than *bioethics* before 'problematisation' and 'boundary-making', as the introductions to these concepts prevent a normative interpretation. In short, I do not use the concept of 'bioethical' to pass value judgement unless explicitly indicated.

Notes on (Co-)authors of the Case Studies

Jyotsna Agnihotri Gupta (Case 2) is an assistant professor in gender and diversity at the University of Humanist Studies in Utrecht, the Netherlands. She is the author of *New Reproductive Technologies, Women's Health and Autonomy: Freedom or Dependency?* (2000) and has contributed to many international publications. Her research concentrates on reproductive technologies and health issues from the perspectives of gender and diversity.

Seyoung Hwang (Case 14) is a senior researcher at the Center for Educational Research at Seoul National University, South Korea. Between 2008 and 2010 she participated in the Economic and Social Research Council project 'International Science and Bioethics Collaborations' as a research fellow at the University of Sussex, UK. Her research interests include public understanding of science and bioethics education.

Masae Kato (Cases 3 and 6) is a postdoctoral researcher at the University of Amsterdam. She conducts research for the project 'Dutchness in Genes and Genealogy.' Observing collaborations between archaeologists, geneticists and genealogists, she observes how knowledge interactions contribute to notions of Dutchness in the Netherlands. She is the author of *Women's Rights? The Politics of Eugenic Abortion in Modern Japan* (2004)

Prasanna Kumar Patra (Cases 4, 7, 16 and 17) is Reader in Anthropology at Utkal University, India, and Postdoctoral Fellow at the Centre for Bionetworking, University of Sussex, UK. His current research for the project 'Bionetworking in Asia' examines the complexities in proliferation, translational bionetworking, and collaborations of stem cell research and therapy in India. He publishes books and articles on Indian tribal areas, genomics, biobanking, and stem cell research in India and Japan.

Suli Sui (Case 5) is an associate professor of the social science department at Peking Union Medical College. She holds a PhD in Social Sciences from Amsterdam University and a Masters of Law from Renmin University, Beijing. Her interdisciplinary research focuses on the crossroad of medical science, law/bioethics and society. She publishes widely on genetic testing, medical law and stem cell research in both Chinese and English international academic journals.

1
Introduction: From Global Moral Economy to Assemblages of Life

What is biotechnology about? Simply put, biotechnology puts knowledge of life, or life itself, to use in order to sustain, repair or enhance life. The United Nations Convention on Biological Diversity[1] defines biotechnology as 'Any technological application that uses biological systems, living organisms, or derivatives thereof, to make or modify products or processes for specific use'. Biotechnology harbours life-changing potential to cure the diseases of many patients, and promises evolutionary enhancement of the human species, to the delight of some (Harris, 2011). However, advanced biotechnology, including genetic engineering, tissue engineering, genomics and stem cell research, is controversial, and not only for the hope it has sparked in—and the disappointment it has brought to—patients and their families. It is controversial also for its potential to radically change humankind and its environment on this globe (and beyond), for the financial and human resources it absorbs and for the uncertainty that surrounds it owing to its experimental nature. 'Bioethical' guidelines have been drafted to facilitate consensus about research conduct and to bolster its development, while trying to prevent and minimise the occurrence of ethical problems and allegations. In many ways, these guidelines are experimental, especially as they tend to vary internationally in their wording, spirit and application.

Bioethics as an academic discipline; however, is overburdened, taking on discussion that requires far more scientific expertise and human experience than can be offered by the relatively limited number of scholars with a background in the humanities, law and medicine involved in its development. Like demography and epidemiology, bioethics addresses questions related to issues of life and death with moral and practical implications for the wider society, including the running of hospitals, life science research, the pharmaceutical industry, and policies on public health. While bioethics plays a key role in the regulation and justification of the life sciences, as implicated above, at the same time it is shaped and conditioned by what I introduce here as 'life assemblages'. The concept of life assemblage as a heuristic methodological tool serves to define communities that share questions related to the definition of what is 'a life worth living'. It also refers to the material, cultural and political conditions of communities that share notions and discourses about how to sustain a life worth living and about fundamental issues of whether and how biomedical science and technology should serve, maintain, repair, improve, enhance and extend life. A life assemblage develops under particular socio-economic conditions, shifting boundaries of knowledge, changing conceptualisations of the body, and forms of political organisation that underlie activities of assembling and reassembling life through technological interventions. A life assemblage shares questions about the biomedical intervention in the lives of citizens, the storage of human tissues and data on their lives, the kinds of human life given birth to, and notions of what is a normal or natural life, and what is not. Life assemblages are not new in the sense that some of these questions, for example on how to extend life, to a certain extent have been important in societies for many centuries. Thus, a major quest of Chinese Taoists was attainment of longevity (Sivin, 1980; Cooper, 1990). The question of longevity allowed reflection about a limited number of life issues that, in principle, were thought to be resolvable within the Taoist paradigm linking alchemy to notions of society. But, in life assemblages, the social and political boundaries that define moral life are continually in flux. In life assemblages, members share mindsets that assume moral change towards life as inevitable and experience the transgression of ethical boundaries as a normal result of developments in science and technology. Whether members of life assemblages agree or not, assemblages of life are organised around questions of whether or not to sanction life, give life, intervene with life, extend life, transplant life and end life on the basis of scientific knowledge and technological intervention. In life assemblages, rather than making decisions to find a cure for a particular disease, store blood for emergency situations or protect a community from a contagious disease, life science communities, including pharmaceutical industries, hospitals, charities, regulators and 'bioethicists', collectively engage in activities that focus on acquiring knowledge about life, repairing life, regenerating life, combining life and justifying life interventions on the basis of ethical principles, guidelines and decisions. Such 'rules' are developed by various social groups, including national ethics committees, regulators, policy-makers, and international organisations, and implemented in interaction with life assemblages. Although the legitimacy and representativity of bioethical rules are usually contested, members of life assemblages share a direct or indirect preoccupation with the discourse around and the questions of whether, to what extent, and how to assemble and lead life.

In life assemblages, socio-political institutions continually redefine and integrate the inputs of religious, cultural and spiritual traditions with more formal bioethical notions to guide life science and biomedical research. On the one hand, this allows life assemblages to become increasingly complex, varied and specialised, hampering communication between life assemblages. On the other hand, the discussion on norms and values within life assemblages increasingly concentrates on and interconnects issues of birth, reproduction, life and death together. Thus, questions concerning the relation between birth and the productivity of populations, the utilisation of the death of one to save the life of another, and the transplantation of stored tissues and organs have become part of a discussion linking debates about euthanasia, eugenics, regenerative medicine and stratified medicine. These themes are subject to moral deliberation and regulation in societies, and are analysed and systematised as approaches in the relatively young disciplinary field of bioethics. Although it is not clear what and whom exactly these bioethical approaches represent, they have become part of national, regional and global life assemblages. For the boundaries of moral traditions and religious taboos are shifting, not so much by dint of shared decision-making informed by religious, social and cultural values, but because a worldwide industry of life science research and biomedical business activities produce new life potentials that have overtaken what societies thus far have imagined possible. Scientific and industrial innovation in the life sciences, then, has led to the adoption of regulatory and bioethical measures that societies still have to come to terms with.

Bioethical Capacity Building

International co-operation in the global biomedical economy links the collaborative efforts of scientists and medical professionals between many countries. But the role of science, technology and bioethics in the creation and transfer of new knowledge and wealth has been typically framed by high-income country policy-makers. Focusing on Asia, this book investigates some of the practical and conceptual implications of bioethical capacity building in genomics, biobanking, genetic testing and in stem cell research in Asian societies. Bioethical capacity building is generally thought to be important to the development of the life sciences. First, it has protective potential. For instance, it aims to protect patients and donors of human tissues through bioethical guidelines on informed consent and experimentation. Bioethical guidelines can also protect researchers from litigation, and from allegations of immoral research. Second, it provides a mode of stability in a risky field of science innovation. Thus, for funders of research innovation, the presence of bioethics committees and guidelines serves to guide or prevent research that is accompanied by a high risk to patients, researchers, reputation, and the financial or academic success of projects. Third, the existence of bioethical capacity can elicit trust from the public, people potentially involved in the research, and support discussion and the acceptability of new developments in the life sciences.

In a global context, however, bioethical capacity develops amidst the existence of diverging standards of wealth, diverging cultures and diverse political preferences. Bioethics, then, may be of diverging significance to patients and researchers, and play a different role in the implementation of research at different locations. This book aims to contribute insights into the different drivers of commercial and academic research institutions and forms of bioethical capacity building by state and non-governmental organisations, while exploring the ways in which moral discourses of development and economic growth are framed and, increasingly, reframed through government policies in Asian countries and the need for international collaboration. Bioethics as shaped in the context of a life assemblage is contested. Thus, 'bioethics' in European and American socio-legal discourses is seen both as a form of 'human rights' (Ashcroft, 2010) and as 'dubious' (Fiester, 2013). In this volume, the emerging voices of bioethics in Asian societies are central, and we trace how these are formulated, represented and contested in the management of life science discourses across societies.

Discourses on Asian societies portray Asia in public discourses variously as backward and advanced, wild and civilised, authoritarian and democratic (Dennis, 2002; Sleeboom, 2004). The appearance of both technological bandwagons and individual pioneers in Asia has complicated the image of recent life science developments in that part of the world. Asia has become dotted with cutting-edge life science hubs, such as Biopolis in Singapore, Penang Biotek Park in Malaysia, Biotec Centre in Thailand and many centres in the metropolises of Asia, including Delhi, Bombay and Bangalore in India, Seoul in South Korea, Tsukuba, Osaka and Kyoto in Japan, Shenzhen, Beijing, Shanghai, Guangzhou and Tianjin in China, and Taipei in Taiwan. Newspapers report intrigues around 'rogue' stem cell therapies developed by Guita Shroff's NuTech MediWorld in India, Huang Hongyun's Western Hills Hospital in Beijing and Tianjin, Beike company in Shenzhen, BCRO in Indonesia, and Restoring Hope and Theravitae in Thailand; and the media has fully exploited the scandal around the fabrication of science data in South Korea by the team of Hwang Woo-suk and the reportedly nefarious stem cell research practices in China—all in all resulting in a dubious image of some science efforts in Asia. At the same time, we hear about the soaring numbers of international peer-reviewed science publications in India and China (*New York Times*, 2010; The Royal Society, 2011), successful cloning efforts by the reviled Hwang Woo-suk team and the pioneering induced pluripotent stem cell (iPS) developments by Yamanaka at Kyoto University in Japan. Moreover, we hear about the successful sequencing results in genomics by Beijing Genome Institute in Beijing and Shenzhen, the establishment of large-scale biobanks in Delhi, Taipei, Tsukuba, Singapore, Taizhou and Kunming, and the success of pharmaceutical conglomerates in India, China and Japan. While the number of scientists and companies keen to collaborate with counterparts in Asia is increasing, uncertainty reigns about the ethics of life science research and applications. This uncertainty has led to questions about the development of ethics in life science ventures and collaborations among researchers from diverging regulatory backgrounds, both within and between countries.

Bioethics and Global Moral Economies

A politicised view of cultural values assumes that ethical values can be negotiated, compromised in exchange for political resources. And, if a currency has international credibility, is dependable and generally convertible, it can facilitate the exchange of values, and effect a change in political positions and the redistribution of political resources. This notion of global moral economy, discussed in Brian and Charlotte Salter's article 'Bioethics and the Global Moral Economy' (2007), shows the important role allocated to bioethics in the global moral economy: it provides an apparent neutral currency with which cultural values can be measured, positions priced, and deals arranged. Indeed, without an active moral economy, the economic viability of the life sciences is though to be severely constrained (Salter and Salter, 2007: 559, 561). I use this political notion of global moral economy here as foil against the concept of life assemblages, which emphasises factors significant to the locality in which life sciences and bioethics develop and are applied. Bioethics, referred to above as an apparent neutral currency of the global moral economy, is more popularly described as 'the study of typically controversial ethics brought about by advances in biology and medicine' (http://en.wikipedia.org/wiki/Bioethics). As the question of what is controversial ethics largely depends on whose voices are heard, as implied above, discussions on bioethics have to be understood in the context of the kind of media, public debate, academic system, civil society, regulation, culture, life science and biotechnology that characterise a community.

Even though some awareness exists that bioethics is variously regulated and implemented variously in countries with diverging political and legal regimes (Jasanoff, 2005), it is still assumed that a trading in life values 'is possible because there is a shared understanding of both the value of the political resources involved and the rules of the game and a common acceptance of the need for political movement' (Salter and Salter, 2007: 559). Although counter-narratives are incorporated within the prevailing body of bioethics through the construction of 'legitimate differences', outright opposition to the bioethical search for compromise dominant players see as a disqualification for which the penalty is marginalisation, if not ostracisation, from the regulatory process (Salter and Salter, 2007: 561–2).

As internationally prevailing bioethics has been framed in the context of traditionally dominant life science and health institutions mostly located in the USA and Europe, the question arises of how biotechnologies in Asia develop in the sociocultural and political contexts of moral economies there. In the twenty-first century, also in Asia, large investments into the infrastructure and dissemination of the life sciences are propagated and justified by promises of innovative technological applications in regenerative medicine, genetic diagnosis and repair, and the advancement of knowledge in fields ranging from molecular biology to translational research. But the application of new biomedical technologies in societies has led to the identification of a host of social, psychological, religious and financial problems. Initially, solutions to these 'bioethical' problems were explored mainly through the disciplines of philosophy and law. In the USA and Europe, following the funding of genomics research initiatives in the 1990s, major sums were invested in the study of ethical, legal, social and institutional aspects (ELSI) of genomics and regenerative medicine. But, gradually, the role of the social sciences in this 'novel' area gained importance. The initial emphasis on the legal and philosophical aspects of the life sciences was a result of the need to regulate and protect the work of scientists, to facilitate the progress of such work, and to reassure the public of its safety and protection (Fox and Swazey, 2008). When it became clear that public views of the life sciences and their applications in society were going to have an impact on science funding, innovation and technological advancement, the demand for social science studies increased.

Bioethical Governance and Global Assemblages

Initially, social studies of the life sciences were concentrated in relatively wealthy countries with long-established biomedical disciplines, such as the USA, countries in Western Europe, and Australia, New Zealand and Japan. But soon interest grew in investing and collaborating with Asian countries that were starting to invest heavily in the life sciences, including China and India. Reasons for collaboration in the areas of genomics and biobanking ranged from the presumed availability in these countries of unique, ethnically isolated and medically naïve populations to the existence—at the time—of less stringent regulatory regimes (Pomfret and Nelson, 2000; Sleeboom, 2005). The exploitation of such regulatory 'bioethical vacuums' (Sleeboom-Faulkner and Patra, 2008) did not last long once it was recognised that research could only flourish if recognised as ethical. And it is especially bioethical governance and the ELSI aspects of the life sciences that are thought to be of relevance to current research and discussion platforms on the life sciences in Asia (Bionet; Eubios; Ethox; SMAP).[2] An important question here is whose ethics come to represent official ethics, and how. For although the research involves countries, interest groups and markets across continents, the terms and discourses of bioethical governance have been largely predefined in the context of life science development in a few wealthy countries. Political, academic and social institutional histories have been largely ignored in the standardisation of bioethical guidelines and in the assessment of the bioethical capacity in low- and middle-income countries: the ability to adapt to 'international' standards and bioethics institutions without major investment has been uncritically assumed. In the meantime, a major concern for scientists in collaborative life science initiatives remains the availability of 'international' standards of bioethics and safety to safeguard the quality of research and its ethical applications (Harris, 2002; Hennig, 2009; Hill, 2009; Kiatpongsan and Sipp, 2008).

Following in the footsteps of lawyers and philosophers, social scientists and bioethicists have been asking how bioethical governance can be introduced to governments, in hospitals and in science institutions (e.g. Bionet Report 2010; Ethox ISSCR; SNAP). Most Asian countries have followed examples from the USA and Europe in setting up bioethics institutions to facilitate science research. However, in most Asian countries, the social sciences have been largely excluded from the process. To better understand the relation between life science, bioethics and society in Asian countries, this book examines the 'situational ethics' or politico-economic circumstances (Ong and Chen, 2010: 13–14) in which regulatory activities (moulded via modern nation-state conditions of regulation, healthcare, policy-making and cultural cosmologies) develop as a result of advancements in the life sciences across modern nation-state boundaries. The reason for studying Asian life assemblages lies in the need to take seriously the interaction of what is regarded as dominant 'global' moral economy in the life sciences with 'local' moral economies in Asia. The fact that we can think of bioethical guidelines in terms of the 'global' here is indicative of the marginal position of the bioethics of late-coming 'local' economies. What is needed, then, is to show how the multiplicity of bioethics in Asia appears to be 'localised' as a result of a process that is interpreted as the 'globalisation' of life science research and biotech applications in transnational settings. In this process, as explained below, transnational competition and collaboration in the life sciences involve the development of diverging life assemblages.

The global articulation of bioeconomies and bioethics has been described from various perspectives: as based on global free-market competition, where scientists and companies are in competition with each other (including through multiple forms of collaboration, collusion and exchanges); and as based on broad neoliberal principles, where nation-states and political federations (such as the European Union, Association of Southeast Asian Nations, etc.) facilitate and back up the competitiveness of their life science industries (Gottweis et al., 2009). Transnational developments in the life sciences have also been analysed from the point of view of infrastructural and historical inequalities (Sunder Rajan, 2006). From this perspective, 'free market' competition in the life sciences builds on institutional, regulatory, socio-political, biological and health inequalities between geographical regions and nation-states. If we hope to understand the transnational dynamics of biotech developments, then free-market competition, neoliberal and protectionist nation-state policies and institutional developments (including economic politics, cultural traditions, social organisations and material resources) need to be analysed together as a global assemblage (Ong and Collier, 2005).

The concept of 'global assemblage' refers to 'projects of various kinds that have global reach and which are refracted through particular localities as scaled and structured versions of the dominant paradigm' (Ong and Collier, 2005: 4). According to Ong and Collier (2005) global assemblages are intrinsic to the process of globalisation and the capacity for similar ideas and practices to manifest in a wide variety of locales and work by very different rationales. The articulation between locales occurs through entities such as international policies, ratifications and standards that are drafted by powerful organisations and which are taken up in different contexts for different reasons. The idea that similar practices develop as part of widespread life science practices underlies the use of the term assemblages here. For, based on similar ideas, plans and practices, the many life sciences hubs dispersed over the globe are also part and parcel of different locales with their own governance mechanism, and socio-cultural and economic contexts. The question here is whether 'local' versions of bioethics are just expressions of subjugation to 'global' pressures or whether they are a product of creative agency and strategies of adaptation that actually help to mould what is often regarded as 'global', 'Western' or 'first world' ethics.

In my view, global assemblages are about both competition and collaboration, where forces and activities between units at different organisational levels (economic management, political regulation, national support) both clash and reinforce one another, so that, for instance, international collaboration between one unit (e.g. universities or companies) can augment the competition between other units (e.g. nation-states). In some cases, global corporate competition between nation-states can be about varying the *very rules* that give competition meaning in wealthy societies (Petryna, 2005). In the life sciences, these rules may relate to the regulation of the life sciences, bioethics and experimentation, including regulatory discrepancies, differences in standards of living, differences in health standards and healthcare provision, differences in the treatment of marginalised human groups and differences in the protection of humans, natural resources and the natural environment to further economic and strategic interests. In other words, alongside internationally accepted rules of competition and nation-state protectionism, there exists resistance to, and pressures against, the accepted rules and conventions of competition and collaboration, leading in some cases to the creative exploitation of regulatory gaps between and loopholes within countries (Sleeboom-Faulkner, 2013a). Using case studies and empirical fieldwork in China, Japan, India and other Asian countries, this book illustrates, compares and analyses such issues of equality, ethics, governance and policy-making regarding life science enterprises from national, regional and transnational perspectives.

Life Assemblages

An exploration of the moral trajectories of the life sciences and the development of bioethics institutions in Asian contexts can yield insight into the interplay between the diversity of 'local' moral orders and tenets that have gained international currency. On the one hand, developments in the life sciences are defined in terms of the need for scientific knowledge and technological innovation in support of human health; on the other hand, the life sciences are driven by economic interests and political and legal institutional structures that condition the development and organisation of innovations that may lead to successful human health interventions. As understood in this volume, life assemblages incorporate shifting and contradictive ties between commercial interests in and moral drivers of the life sciences. Such assemblages are networks of human and non-human elements (Latour, 2005) that underlie developments in ethics and decision-making. The identification of life assemblages at the conjunctions of the life sciences and bioethics in Asian societies can help clarify and explain the persistence of 'local' notions of life science governance.

Life assemblages are characterised by particular modes of combining sources of ethical authority with the advance of life science practices and applications. Concepts of ethics buttressing life assemblages include notions of fairness, aid, charity, healing, inequality, exploitation, negligence and abuse, and inform healthcare provision, public health, scientific funding and bioethics institutions. The notion of life assemblage complements that of global assemblage (Ong and Collier, 2005). When applied in the life sciences, a global assemblage combines factors from the global socio-economic environment directed at producing science innovation. A life assemblage, however, reflects the resultant practice of multiple sets of rules and principles of distribution, varying from free market principles of competition to collaborative principles of 'benefit-sharing', altruistic principles of financial and biological donation, bureaucratic rules and regulation, and traditional and religious values to produce the ethical conditions through which the life sciences develop. Life assemblages, then, follow a different logic (if any) from the sets of formal bioethical standards developed to facilitate innovation in the life sciences and to protect patients and public morality. For life assemblages express the various ways in which both gains and costs of developments in life science practices result through the mediation of society, including its healthcare access and economic inequalities, social injustices, scientific and religious ideologies, and business rationale. In short, life assemblages inform us about how life sciences and bioethics are translated and embedded into the societies in which they are used, shaped through relations of power and agency, and function as part of scientific development and relational networks.

As a heuristic tool, the construction of life assemblages identifies economic and moral drivers of developments in the life sciences that interact to produce research policies and bioethics regimes. It aims to capture the real-world heterogeneity of the things, rules and actors that go into the making of bioethics and research regulation. It does this by tracing the connections between actors, institutions, knowledge, values, facts and practices. I here follow Bruno Latour's view of an assemblage: items in an assemblage are connected to other elements, and in affecting other actors produce identifiable effects (2005: 63–86). Such elements also include things that, at face value, have little to do with the life sciences or research ethics, but still influence research policy outcomes. Thus, the elements of 'rural education', 'feminism' and 'patriotism' can be crucial to the development of the life science in a particular life assemblage, while playing a marginal role in others. Similarly, research policy may be affected by things happening far away, such as the scientific discoveries or changes in overseas regulatory regimes. Thus, the discovery of iPS led to a new constellation of research practices and bioethics discourses in many countries, and regulatory change of human embryonic stem cell research (hESR) under former US president George W. Bush set in motion change in the regulatory regimes and stem cell research policies in Asian countries.

But seemingly unrelated things and connections hardly make for a convincing argument for their significance as a life assemblage. If, at first sight, connected elements appear to have little to do with an assemblage (being, e.g., of global provenance), how can one even begin to define a life assemblage? After all, developments in the life sciences in different world locations, the religious context in which scientists work, traditions of holding public debate, the availability of research funding, the sociopolitical history of science institutions, the social relevance of particular research innovations and the political clout of a particular government may all be relevant to the development of regulation and bioethics. It is exactly because these components of life assemblages are elusive and context-dependent that their identification and delineation requires ethnographic research and the explicit tracing of interregional and transnational connections. The connection between these elements needs to be explicated and explained in the light of trends revealed.

The notion of life assemblage, with its various dimensions of power, agency, bioethical translation and embedding, risk signatures and bionetworking—as explained in this volume—can only facilitate limited comparative analysis. Such limited comparative analysis allows the understanding of why and how a particular technology is used differently in diverging circumstances. For instance, a comparison of genetic testing in the context of life assemblages could be useful in that it facilitates the creation of hypotheses about the frameworks in which the people undergoing genetic tests make choices. In the case of hESR, comparative analysis of life assemblages can generate hypotheses, not just about religious or traditional values prevailing in certain regions, but also on the different ways in which countries regulate life sciences research, the embedding of values in the work of scientists in particular areas, and the modes in which life issues are subjected to debate in society. The way particular practices connect to life assemblages provides insight into agency, wealth, ideology, politics, and other forms of power. But, as the units of comparison depend on the place of power and agency in a life assemblage, which is linked to the problem we are investigating, the research problem and life assemblage need to be understood together. This makes for a poor basis for systematic comparison. Nevertheless, comparison raises critical potential and generates ideas for further research. If we are interested in the relation between family size, abortion and ethics, for instance, constructing life assemblages can be of great use. When we discover that state power in one country is more important in the determination of reproductive choice than in another country, where, say, the family clan is of greater importance, then we know that interviewing couples about the ethics of abortion alone will not be sufficient to reach our goal of understanding the role of ethics and abortion in family size. It tells us that we need to explore more fully the research problem-relevant connections of the life assemblage. Comparison here can provide us with hints (that we can use to make hypotheses) as to what kind of connections may be of significance in solving our research problem. By alternately using the research methods of comparing life assemblages and the identification of interlinking elements of life assemblages, we can extend our understanding of how life sciences relate to society.

The concept of life assemblage especially serves to raise awareness of how life science issues are configured in robust contexts that cannot be presumed to be amenable to what we think of as universal moral values. It is a methodological construct to draw attention to the embedding of moral economies. The relevance of the concept stands or falls with the way it is utilised in the analysis of research problems. Defining the boundaries of a life assemblage relies on the connections we identify in the context of a certain research problem, and how we identify and analyse elements as being connected or linked with other life assemblages. Attracting attention to the contextual importance of the life sciences, its regulation and bioethics does not, of course, mean that economic theories do not apply. These, however, need to be understood through the way economic markets are embedded as fundamental dimensions of life assemblages.

From a Critique of Neoliberalism to the Delineation of Life Assemblages

Recently, a body of literature (Hayden, 2003, 2007; Sunder Rajan, 2006; Helmreich, 2007, 2008; Cooper, 2008) has analysed the relationship between biological reproduction and capital accumulation in a world many academic authors characterise as neoliberal. Anthropologist Kaushik Sunder Rajan first identified the promissory nature of speculative biocapital in the life sciences as a new phase of capitalism. Productive and reproductive life together produce value and remain the bases of capitalist accumulation (Cooper, 2008: 7). By concentrating especially on economic values, these works aim to explain economic and political drivers of developments in the life sciences. While life science investments based on the belief in the success of the life sciences to generate wealth illustrate how forms of capitalism have led to new forms of commodification (including the selling of parts of the body and body tissues) and the commercialisation of reproductive labour (Cooper, 2008: 101) and the creation of 'biovalue' (Waldby and Mitchell, 2006). But the ways in which the applications of new biotechnologies are experienced and evaluated in countries with diverging economic and political systems and sociocultural realities has received little attention. For this reason, and as argued in Andrew Kipnis' (2007) critique of the ambiguous and abstract use of the notions of neoliberalism, there is a great need to be more specific about the contextual embedding of the economic and political systems academics write about. For instance, political scientist Melinda Cooper illustrates how 'neoliberalism' reworks the value of life as established in the welfare state and the New Deal model of social reproduction. Cooper argues that these have tended to wear away 'the boundaries between production and reproduction, labour and life, the market and living tissues' (2008: 9). But no matter how salient these trends seem to be in the light of the enormous investments in the life sciences and in healthcare, the variety in the reproduction of values in low- and middle-income countries, and countries with different juridical and regulatory systems remains underexposed. And, as I shall argue here, it is especially the differences that remain hidden under the generalisations about world systems that may form the impetus for thousands of actors to move and interact in unpredictable directions, and incite major social changes that we have not noticed as yet. For instance, economic theory alone cannot explain why human embryonic stem cell scientists report that Japan suffers from regulatory suffocation, while it is (for as yet) also a freehaven for experimental stem cell therapies (*Nature* Editorial, 2013: 5). Neither are there universal 'neoliberal' principles that can explain the important changes in how the relations between things, life and society are valued and deployed in the life sciences, which is why we cannot ignore life assemblages. Especially when it comes to the question of how everyday values alter, hamper and resist life science and bioethical practices, it is clear that little is known about the socio-political relations and cultural values that shape scientific activity. Using comparative and empirical research, the delineation and comparison of life assemblages will assist in the identification of the roles of socio-political relations and cultural values of life behind life science developments in Asia.

In this book there are many references to 'Asia' and other regions, which needs clarification. When referring to Asia, East Asia and Europe, I point to the geographical locations in general without further associating these with a particular culture, religion, economy or political system unless so indicated. For it is exactly the function of life assemblages to define the most suitable units of analyses in the examination of a certain problem field. If national regulation is relevant to a research problem, then it is likely that the nation-state is relevant as one of the units of analysis. But, as will be shown throughout the chapters, what and which units of analysis are relevant in life assemblages cannot be said before having investigated a particular research problem in practice. Thus, apart from nation-states, other geographical units, regions, metropoles, cities, towns and cultural units, such as the family household, the nuclear family, Buddhist culture and tribal communities, may be of crucial importance to the analysis. Amidst such uncertainty about what are relevant units of analysis, comparative analysis is not straightforward, and is mainly useful as a source for generating questions about the similarities and differences between the units that we interlink. Such questions sensitise us to the contextual factors important to life assemblages. Life assemblages, then, are understood, first, by tracing connections, and, second, by comparing the units that we find are connected. This might generate new hypotheses about the similarities and differences between life science practices. Thus, when we find a link between a Chinese biotech company and an Indian hospital, we might gain new insights by comparing the conditions under which Chinese and hospitals of a similar nature operate. In this way, the methods of comparison and tracing links form the basis of the analysis of life assemblages as described in this book.

Overview

A constellation of social science issues is central to this book, including population health, biopower, human biobanking, public debate, risk, human experimentation, inequality and stigma. A discussion of these issues aims to shed light on the various dimensions of 'life assemblages'.

Intended to provide the reader with insight into the ways in which the relations between bioethics and the life sciences are played out in multiple Asian life science practices, the book is divided into chapters containing examples based on empirical studies in various part of Asia. Each chapter starts with a short background explanation of the main terms and arguments discussed within it, linking it to a particular dimension of life assemblages: reproductive agency, Chapter 2; biopower and genetic screening, Chapter 3; the translation of bioethics in biobanking, Chapter 4; problematisation and boundary-making in the bioethics of hESR, Chapter 5; forms of risk in stem cell research, Chapter 6; bionetworking and science collaborations in stem cell research and therapy, Chapter 7. The introduction of each chapter is followed by two or three short case studies, grounding the general discussion in empirical reality. Each chapter closes with a reflection on the issues raised in the case studies. The concluding chapter, Chapter 8, revisits these reflections in a discussion of life assemblages.

Each chapter relates controversial issues to debates important in the social sciences. The debates themselves are linked, in turn, to the various dimensions of life assemblages. Chapter 2, entitled 'Reproduction and the Question of Eugenics: Prenatal and Premarital Testing in China, India and Japan', raises the question of whether technological intervention in birth planning can be regarded as a form of eugenics, and, if not, how we are to understand forms of birth intervention through new reproductive technologies in different socio-economic, political and religious environments. This chapter sheds light on the field of reproductive decision-making, focusing, in particular, on the interaction between infrastructural aspects of social reproduction and agents of decision-making. Adopting reproductive technologies used in state population policies, the question central to this chapter is to what extent individuals and social groups are driven to use them, regard them as an opportunity to effectively change their lives, or use them to create a space for rebellion against, say, state or community policies and ideologies.

The concept of state eugenics in Western societies is historically associated with the systematic elimination of groups of people based on racial concepts and ideologies of population enhancement. Awareness of the atrocities wrought in the name of eugenics seems to have influenced the attitudes of prospective parents towards the question of aborting fetuses affected by disabilities. In some countries, 'eugenic abortion' can be regarded as wrong, while in others it is viewed as a valuable means of averting the disaster of raising an 'affected child'. The three case studies in Chapter 2 discuss this issue by examining the ways in which prospective parents and couples make reproductive decisions when confronted with undesired reproductive information. Case 1 looks at compulsory and voluntary birth planning in urban and rural environments in China, and the choices people face; Case 2, authored by Jyotsna Agnihotri Gupta, examines the considerations important in the decisions of families to abort 'children' with potentially unwelcome traits during prenatal and genetic counselling sessions in clinics; Case 3, co-written with Masae Kato, examines the narratives of parents that have and have not aborted fetuses tested to be positive for unwelcome syndromes, and the latter group's reasons to give birth to and raise an 'imperfect' child. Chapter 2 questions the ways in which the concept of eugenics is traditionally applied as a nation-state ideology or as a new form of individual decision-making (Taussig et al., 2003). It draws attention to the necessity of keeping an open mind as to how the decisions made by individuals are as much decisions made through other social units of organisation, including couples, families, households, communities, clans and the state. Moreover, the choices that are made are not just based on the choices and interests of these social actors, but are also framed by infrastructures of regulation, kinship, wealth, education, healthcare access, technological capacity and dominant life values. The chapter shows how various socio-economic, cultural and political elements form diverging life assemblages in the field of reproduction. They indicate that national policies are of great influence on local paradigms of bioethics, but that this influence differs starkly, and often crucially, in life science practices. In what ways, and how agency affects frameworks of reproductive choice is discussed in the chapter's conclusion.

When adults are tested for certain genetic traits or syndromes, this is called genetic carrier testing, and it results in forms of genetic knowledge. Chapter 3, entitled 'Biopower and Genetic Group Ideologies: Genetic

Carrier Testing in India, China and Japan', is about the availability of genetic testing technologies, the application of genetic knowledge and the way communities deal with new choices regarding the intervention with what is viewed as natural life. The chapter discusses Foucauldian notions of biopower and biobureaucracies in the context of genetic carrier testing and screening in relation to issues of individual autonomy, free choice and the moral economy of development in some Asian countries. It has been theorised in a European context that, in using these testing technologies, populations aim to maximise their life values, facilitated by effective biobureaucratic institutions through which biopower is exercised productively and positively. The three case studies in Chapter 3 comment on this conceptual use of biopower as (1) characterised by biobureaucratic institutions that productively exercise biopower (Case 4); (2) functioning through biobureaucracies that are conducive to autonomous individual choice (Case 5); and (3), as conducive to the rationalisation of society and individual choice (Case 6). Case 4, co-authored with Prasanna Kumar Patra, shows how, in the absence of biobureaucratic institutions, screening technologies for sickle cell anaemia in tribal areas in India are attributed with quite different powers and meanings; Case 5, co-authored with Suli Sui, examines how the effects of screening for Duchenne muscular dystrophy in mainland China are subject to state policies and environments that are not conducive to autonomous individual decision-making; Case 6, co-authored with Masae Kato, shows how the cultural meaning attached to genetic disease affects the way in which genetic information can be used for 'rational' purposes. The case studies make clear how the power effects of new genetic technologies do not automatically generate institutions conducive to autonomous choice amenable to political decision-making, as decisions about genetic testing are conditioned by diverging frameworks of choice. Together, these case studies show the necessity of an alternative conceptualisation of the effect of applying genetic technologies in society. The life assemblages described here do not confirm the view that the use of genetic technologies rationalise human life in society. How genetic technology affects society (biopower)—apart from being used as a source of genetic knowledge and medical decision-making— depends on the infrastructural frameworks that allow genetic screening to be a vehicle used for social subjugation and discrimination, a source of saviour and emancipation, and a statement of critique directed against the rationalisation of life.

The management of public and individual healthcare has come to depend on technologically advanced systems for maintaining healthcare data (and blood and tissues in biobanks), requiring new methods of data collection, maintenance and management. Difference in opinion exists in a body of bioethical literature on the ways in which personal and confidential data should be collected and stored, and who benefits from the data collections. Chapter 4, entitled 'Human Genetic Biobanking and Life Assemblages in Asia: Transnational Moral Economies of Health, Progress and Exploitation', examines how moral economies of biobanking on global, national, local and corporate levels interact. It raises questions about the kinds of bioethics institutions deploy while collecting medical data and tissues in biobanks, and their application in different societal contexts. Of central concern here is the question of what happens when scientists in Asian countries are confronted with pre-existing 'international consensus' about bioethical procedures, but have neither the financial and bioethical capacity nor the political clout to apply them. Case 7, co-authored with Prasanna Kumar Patra, examines the efforts made to set up the Permanent National Repository for Human Genetic Resources and Data of India in the light of international regulation, and the reasons for its eventual abandonment; Case 8 depicts the influence of patriotic politics involved in setting up a human genetic biobank in Indonesia and the related bioethical regulation; and Case 9 looks at the role of biobanking in the process of nation-state building in mainland China and the importance of bioethically-correct international speak. The case studies in Chapter 4 clearly show that bioethical capacity is not the same as bioethical knowledge. What we learn, in fact, is that those lacking bioethical capacity need to have detailed knowledge of bioethics institutions in order to survive professionally. We see, then, that a life assemblage incorporates the connections between elements that form the bioethical practice of biobanking in a particular context, but may not be the same as the bioethics advocated in formal terms. The divergence in life assemblages opens up a playing field in which 'standardised bioethics' is not just a source of domination, but also a source of creative manipulation, and contain a critique of both local and general practices of biobanking.

   Bioethics and the way it has developed and spread around the globe has been controversial since its inception. This is so not just because of the existence of stakeholders with clashing interests, but also because bioethics institutions, if adopted from elsewhere, do not comfortably fit with local discourses of morality. Some observers regard bioethics as necessary to developing life-saving technologies, believing that it helps to teach and guarantee respect for the autonomy and choices of the individual. Others regard its principles and aims as instruments of a hegemonic neoliberal world, enabling the exploitation of the vulnerable. Still others regard bioethics as acceptable and useful to guide life science practices as long as it is 'locally' led and developed. Chapter 5, entitled 'Life Assemblages of Human Embryonic Stem Cell Research in China and Japan: Bioethical Problematisations and Bioethical Boundary-making', draws on empirical field studies to question all three views. While focusing on the adoption of bioethics institutions in scientific research centres in China and Japan, the chapter analyses this embedding in interaction with perceived life science developments on a 'global level', which usually means the USA and/or Europe. Case 10 discusses the abovementioned views in light of the development of the bioethics of hESR and cloning in mainland China. Applying the notion of 'bioethical problematisation' to therapeutic cloning in China, Case 11 investigates how demands and pressure from people directly involved in embryo donation (donors, researchers, medical professionals) are connected with and marshalled in support of the politics of scientific innovation. Case 12 uses the notion of 'bioethical boundary-making', a concept that helps indicate differences in socio-cultural and political orientation of Chinese and Japanese stem cell scientists. It shows how global, national and local factors are interwoven and help shape bioethics discourses and practices of bioethics in human reproductive cloning and hESR. The chapter as a whole illustrates how what are thought of as dominant concepts of bioethics are continuously adopted, shaped and reshaped through their application, translation and adaptation into contexts they were not developed in. According to Ong and Collier, 'global forms are articulated in special situation—or territorialised in assemblages, defining new material and discursive materials' (2005: 4). It is exactly when bioethics is territorialised in new contexts that extraordinary effort must be exerted to make it fit locally, an activity that follows pre-existing cultures of adaptation. In Chapter 5, then, we see that such adaptation can have meanings ranging from a purist adoption of bioethics into bureaucratic institutions to its creative absorption in new organisational structures.

The notion that the public should be involved in debating the direction of life science applications has become increasingly widespread. But who constitutes this public, the manner of its involvement and what authority such involvement should have in policy-decisions are questions largely determined by political and participatory mechanisms that vary nationally. Chapter 6, entitled 'Scientists and the Public in East Asian Life Assemblages: Risk, Debate, and the Professionalisation of Bioethics', analyses the views of scientists and policy-makers of bioethics and public debate about hESR in terms of the perceived risk of the bioethics debate to the development of stem cell science and therapies. Using the concept of national risk signature, Case 13 explores scientists' interpretations of bioethical regulation and public discussion, and the effects of regulation and public debate on scientists' careers, institutes and countries. Case 14, co-authored with Seyoung Hwang, on the basis of empirical research among stem cell scientists in South Korea, asks whether the institutionalisation and professionalisation of bioethics should always be welcomed. This question is important with regard to the role we would like civil society to play in science policy-making, and the interaction between scientists and the public. Case 15 examines how the meaning of risk and danger in the context of hESR and iPS in Japan are culturally and politically mediated, and how they change in relation to global developments in these fields. In this chapter, we see that bioethics, if not built on knowledge scrutinised by a broad public, can constitute a risk to scientists for opposite reasons: scientists, in some cases, fear that their work and its ethical aspects may be criticised by 'the people'; but, in others, scientists may worry that they do not have enough opportunity to persuade the public of the benefits of their research, and its safe and efficacious nature. For some, the public deliberation of bioethics is an opportunity; for others, it forms a risk. The difference is not necessarily that some scientists are involved in kosher, and others in rogue, forms of scientific research. In Chapter 6, the notion of life assemblages explains why bioethics is feared or welcomed, and shows the multiple factors involved in its use and evaluation, including pre-existing political mechanisms, general education levels and patient movements.

The political and public mechanisms nation-states have in place for appraising life science research and applications of new biotechnologies are only some of the conditions that shape the development of the life sciences in different countries (Jasanoff, 2005). Transnational connections and collaborations between institutions of a different nature, such as private and public institutions, however, are far more difficult to delineate, as they are based on the actions and calculations of individual networks that operate across the boundaries of spaces characterised by different rules and conditions. Transnational collaborations build on the combination of sociocultural phenomena unique to the political history of particular countries, while private–public collaborations may have to deal with differences in research and service aims, institutional organisation and institutional cultures. And although it is only rare that an institute conducts research purely for the purpose of acquiring scientific results, the perceived differences between scientific and other aims of enterprises and projects can be crucial to those that have a stake in them. Pursuing aims and motives other than those agreed on by collaborative partners may frustrate the expectations of funders, the public and even the collaborative partners themselves. Chapter 7, entitled 'Life Assemblages and Bionetworking: Developments in Experimental Stem Cell Therapies in India and Japan', discusses these factors in the context of the ambivalent situation in which scientists combine entrepreneurial activities with research in stem cell applications. Case 16, co-authored with Prasanna Kumar Patra, explains how bionetworking within India connects patients to a range of therapy-providing institutions, while Case 17 illustrates how transnational bionetworking activities connect scientists, therapy providers and patients in India and Japan. The bioethics that gains currency in such collaborations can only be explained by considering how interests, ideals and facilitating elements are connected within and across territorial and institutional boundaries. Life assemblages, which incorporate territorialised elements of life science research, besides being influenced by nation-state policies, also need to be seen in the light of the dynamics of the assemblages themselves, as they are part of, and linked to, international and transnational developments. Nevertheless, although transboundary developments in bioethics signify new overall developments in the elements constituting life assemblages, territorialised life assemblages are identified by their relative autonomy, stability and integrity vis-à-vis their environment.

The concluding chapter, Chapter 8, entitled 'Reframing the Global Moral Economy: Life Assemblages and Research Objects', links up themes discussed in Chapters 2–7 as dimensions of life assemblages, and shows how they shed light on the conditions in which bioethics institutions in Asia have come about. The concept of life assemblages, further elaborated in Chapter 8, intends to take the discussion beyond the controversy about whether bioethics is a matter of moral economy, capitalist exploitation, free market choice or national sovereignty (Lock and Kauffert, 1998; Knoppers, 2003; Taussig et al., 2003; Salter and Salter, 2007; Sunder Rajan, 2007; Cooper, 2008; Birch and Tyfield, 2013). Owing to historical roots, a life assemblage has a robust territorial, material and cultural identity. So, even under global pressure, life assemblages with their own boundary-making activities, mode of embedding and translation mechanisms, generate their own dynamics. Nevertheless, unequal conditions in production and reproduction within and between countries are propelling forces of innovation and economic investment, paradoxically stimulating both fierce international competition and international collaborations. Despite the presence of these transnational forces, the concept of life assemblages leads us to ask the question of how bioethical guidelines can be sensibly formulated if 'international guidelines' do not respond to local and national needs, aims and conditions. Chapter 8 closes with a discussion of how the 'research object' is relevant to answering this question by exploring how bioethics can become ethical in the eyes of those directly involved in life science research and its applications.

In brief, the chapters of this book explore social and political dimensions in developments of the life sciences and biotechnology that have been controversial in both Asia and the West. Rather than selecting and representing bioethical problems challenging Asian countries, this book explores the different dimensions of life assemblages—agency, power, the translation of bioethics, the embedding of bioethics, risk and public deliberation—through which bioethical 'problems' become manifest. In other words, the book asks questions about the ways newly introduced bioethics have played out in local socio-political, legal and economic institutions underpinning life science research in Asia. Chapter 2 deploys discussions on eugenics and prenatal and premarital testing to ask how certain issues become ethical, and how ethicality is related to the state, the individual and other social units, such as the household, the family and the company; Chapter 3, on genetic testing and screening, asks whether the development of biotechnology inevitably leads to interventions based on rational decision-making and productive use in practice; Chapter 4, on the international pressure to regulate biobanking, asks questions about whose interests and wellbeing stores of data and samples serve, how their research relates to public needs and whether biobanks can survive the requirements of bioethics; Chapter 5 asks how biotechnology and life science become adopted as causes in politics, what kind of politics they serve and whose life science ethics comes to prevail; Chapter 6 asks when and how the public may have a say in what is bioethical and whether the standardisation of bioethics is conducive to civil society in Asia; Chapter 7 explores how the existing multiplicity of bioethics facilitates new research and therapy initiatives in stem cell science; and Chapter 8 integrates the various dimensions of life assemblages discussed in the chapters of this book around the question of public deliberation and life science policy-making.

This book is offered as an effort to explain the robustness of the multiple life worlds and their valued underlying bioethics in a world that is increasingly felt to be subject to never-ending standardisation. The concept of life assemblages subsumes the manifold factors that explain the deployment of particular forms of bioethics more than do the meanings attributed to it in the widely diverging contexts and developments in the life sciences. Governments and scientists interested in more solid bioethical governance may have to involve the people whose support and co-operation they need. For a more stable form of bioethics is attained if its values serve the aims of the life sciences and health institutions in co-ordination with the expectations and values of the people they are aimed to serve. This situation could be obtained by involving those that are directly engaged in shaping bioethics in life science research and healthcare, and all of those concerned about its consequences for themselves and their fellow citizens.

2
Reassembling Populations: Questions of Eugenics in China, India and Japan

Although population health in most countries is a general political concern, population policies that emphasise economic growth pay particular attention to the population's condition to sustain the economy. From this perspective, technology serves to raise the physical and mental health of the population and optimise reproductive success. Much controversy has occurred during the last century about the extent to which intervention in family health and reproduction is justifiable. A main issue here is that of eugenic birth, especially associated with the Nazi practice of racial purification. Over the last decades, eugenic birth has been discussed and analysed in Western countries in relation to flexible eugenics (Lock, 2003) and laissez-faire eugenics, and has been analysed in terms of moral pragmatism (Taussig et al., 2003). Some authors argue that the population policies of some Asian countries make use of new forms of eugenics that resemble the state eugenics of the nineteenth and twentieth centuries, while others deny this vehemently (Chen et al., 1999; Rose, 2007). Thus, questions have arisen about the socio-political implications of reproductive choice in the twenty-first century in relation to individualism and authoritarianism.

This chapter discusses whether new biotechnologies bring more or less reproductive choice to individuals in Asian countries, and how this configuration of reproductive choices reflects the activities of governments, communities, families, couples and individuals on the basis of notions of health, life worth living, family reputation, expenses and burden. This relation between the availability of new reproductive technologies and the frameworks of choice that enable reproductive decision-making is an important constituent of the life assemblage. The former concerns the availability of reproductive technologies in the healthcare provision of various countries, and is determined by factors such as technological know-how, the availability of reagents, the attitude towards prenatal and premarital testing of the particular regime and within a particular country, and the state funding of the healthcare system. The latter, the frameworks of choice from which reproductive decisions are made, refers to the socio-cultural, financial and political factors, including the conditions for access to healthcare, religious dogma, cultural tradition and population policies. Thus, testing for certain diseases without access to healthcare leaves a so-called 'therapeutic gap' (Simpson, 2010); religious ideas about fate and karma may render testing meaningless or sinful; patriarchal traditions that prefer sons may compel women to undergo ultrasound or other forms of prenatal testing (PNT); and, population policies can pressurise couples to undergo premarital testing or to undergo PNT to increase the chances of a 'high quality' child.

The focus of this chapter is PNT. PNT is a technique used to inform couples that are at increased risk of having a child with an inherited 'disorder' about the condition of a fetus or embryo. The diagnosis may help them decide whether to terminate the pregnancy or to prepare for the birth of a child with a disability. Risk factors might include advanced maternal age; a family history of a serious, inheritable medical condition; one or both parents being 'carriers' of mutation(s) in the same gene; abnormal test results, such as ultrasound or first and second trimester prenatal screening tests; and a history of a previous child affected by a serious growth, developmental or health problem. PNT can be carried out to identify specific genetic conditions, including Down syndrome, muscular dystrophies, thalassaemia and haemophilia.

Decisions to give birth to or avoid certain kind of offspring, just as sex selection for non-medical reasons, have to be understood in the light of the life assemblage. For, it is often the pressure to conform to collective values and socio-economic conditions that sets off the desire to have a particular kind of child. Though the pressure to conform might take precedence over the wish to give birth to a child 'negatively marked' by others, individual preference in itself is usually shaped in interaction with collective values of human life. Similarly, the pressure not to have an abortion may be a result of the pressure to conform to the norms of social or religious communities.

There are many intermediate positions that regard abortion as acceptable in some cases, but not in others. In all societies, including secular societies, there seem to be conditions, often related to the age of the fetus, in which abortion is regarded as no longer acceptable (Wertz and Fletcher, 1998). In these cases, testing methods can be used to check the quality of the fetus and consider early-stage abortion. Reasons for opposing the use of testing methods range from respect for the 'human rights' of a, say, 20-week-old fetus and the worry about using invasive testing methods, such as amniocentesis, to fears of the 'slippery slope problem', which presumes that allowing one syndrome to be tested necessarily leads to the testing of other syndromes. This would lead to the selective elimination of future people with undesirable traits, or 'eugenics'. Although the ability to use reproductive selection methods raises fears among some, others regard it as a promising solution to solving problems related to overpopulation, food shortage, expensive healthcare and the extinction of natural resources. It is the possibilities opened up by reproductive technologies to intervene with what is seen in a given society as the 'natural course of life' that has led to a search for scientific ways of controlling the reproduction of communities as a means of reassembling life as desirable and according to plan.

In the context of birth planning, in most European countries 'eugenics' is an important conceptual instrument in warning against genocide and the selective discrimination of certain groups of people, such as ethnic minorities, the disabled or groups targeted for other reasons. The question here is whether certain human traits will become entirely undesirable, who decides about this, and whether society will be cleansed of these 'undesirable' people. Some authors find the use of the concept of eugenics in this context too drastic because of its association with particular Nazi practices, such as ethnic cleansing and insectification (Rose, 2007). Notions of 'state eugenics' cannot adequately capture eugenic developments in the West. Other scholars, however, find a redefined concept of eugenics useful in situations where selective abortion takes place under strong pressure from the local or national community (Mao and Wertz, 1997; Mao, 1998; Nie, 2005). Whether and which concept of eugenics is deployed, then, depends on the kinds of traits undesirable in the newborn—ranging from skin colour and obesity to Down syndrome and muscular dystrophy—and which decision-makers are involved, including individual couples, pregnant women, the family household, the local community and the state.

Attempts have also been made to redefine the concept of 'eugenics' in the light of neo-liberalism and individualism. Neo-liberal eugenics refers to the study and non-coercive use of reproductive and genetic technologies to enhance the biological characteristics and capacities of human beings (Stock, 2003). In a more subtle analysis of its role, Margaret Lock's notion of 'laissez-faire eugenics' implies that the language of eugenics is no longer focused on social policy, the good of the species or even the collective gene pool. Decision-making regarding genetic testing, abortion or participation in genetic screening programmes is dominated by the idea of individual choice (Lock, 2003: 250). And, again, Taussig et al. (2003) have argued that 'flexible eugenics' arises from long-standing biases against atypical bodies, and leads to a tension between genetic normalisation and biotechnological individualism. It is a 'tension between free choice within an array of technically mediated possibilities and in a context of discourses of perfectibility we all live, within dominant ideologies of power' (Taussig et al., 2003: 65–6).

The technological elimination of undesirable offspring is currently mainly practised through pre-implantation genetic diagnosis (PGD) and PNT, while PGD has limited applicability as it is performed only within in vitro fertilisation (IVF) procedures and is hardly an affordable option in low- and middle-income countries (Sui and Sleeboom-Faulkner, 2010a). The question of choosing certain kinds of offspring is more controversial in some countries than in others. For instance, there are no legislative limits on the applications of PGD in the USA, while the Council of Europe's Convention on Human Rights and Biomedicine explicitly forbids its application,[1] and the Dutch government nearly experienced a crisis in June 2008 over whether serious, inheritable forms of breast cancer may be included in PGD. Apart from questioning abortion, the Christenunie (Christian Union) also asked how many disorders could be added to the list of syndromes for which the fetus can be tested and aborted. As for presymptomatic and carrier testing, however, sociologist Nikolas Rose (2007) argues that most Western societies leave it up to the individual, and provide genetic counselling to individuals and their families that find it hard to make a decision. As genetic counselling in most countries is designed to help parents and families to attain their reproductive goals, prenatal diagnosis can hardly be called eugenic in the sense of subject to state coercion. The idea of its non-directive nature and its focus on the individual/family, then, is meant to place it outside most definitions of eugenics associated with the elimination of disabled people or genocide (Kevles, 1985).

Moving away from this theoretical focus on the individual, Latimer (2007) describes what the need for parental choice means where there is a risk of serious defects, pathology and deformity. This *need for choice* is performed not in relation to matters of consumer choice and individual preference, but rather as an aspect of responsible citizenship (Rose and Novas, 2005) or what Latimer calls 'soft' eugenics, where genetic choice is legitimated as rational, reasonable and moral given a family's circumstances (2007: 102). Here, *informed* choice is thought to be in need of 'non-directive genetic counselling'. However, aside from the difficulties of fulfilling the ethical standards of non-directive counselling in practice, the notion of 'non-directive genetic counselling' ignores that the need for choice is itself socially constructed (2007: 102–3). Concepts such as 'non-directive counselling' and 'autonomous decision-making' are similarly relational (Stemerding and Nelis, 2001; Latimer, 2007), and should be seen in the light of what constitutes autonomous choice and in the light of what is socially acceptable choice.

Considering the recent increase of state interference with family planning globally, but particularly in Asia, the question of whether and how this ideal of value-free genetic counselling and autonomous decision-making is practised in Asian societies, some of which have or have had eugenic laws until recently, arises. In Japan, eugenic laws were only abolished in 1996 (Tsuge, 2008). In Taiwan, the Eugenic Health Law of 9 July 1984 stated that physicians advise sterilisation or abortion if necessary to prevent 'genetic, contagious, or mental disease which adversely affects well-being of the offspring' (Wertz, 1999). In the People's Republic of China (PRC) the Law on Maternal and Infant Health Care of 27 October 1994 contained 38 articles devised to improve maternal and child health (Dikötter, 1998; Wertz, 1999). In Singapore, Prime Minister Lee Kuan-Yu's policy statement of August 1983 instructed that the government should encourage college graduates to have more children (Wertz, 1999). Over time, however, these Asian countries have adopted regulation that emphasises the role of the individual in decision-making, using vocabulary such as 'advising', 'encouraging' and 'informing' the patient. It seems, then, that notions of individual decision-making have become more recognised and accepted, at least by governments. A main question here is to what extent the expansion of prenatal and genetic testing in the twenty-first century means that *in practice* individuals are left with choice after diagnosis, and how these choices are framed. A second question is to what extent birth interventions are constituted as 'technological interventions' in contrast with the 'normal' course of life. In other words, to what extent can these interventions be said to be a result of state policies, community pressures, family household decisions or other modes of decision-making, and how and for what reasons do they take place? In brief, what kinds of life assemblages underpin decisions to recreate populations by means of termination?

In the following three case studies of reproductive decision-making in the PRC, India and Japan, the state, the family household and individual couples play significant roles, but the importance of these levels of decision-making differs in the general context of each country. Case 1 examines the concepts of eugenics and premarital testing in the historical and cultural context of the population and one-child policies of the PRC. Even though the policies subjected the population to intervention in reproductive matters on a compulsory basis, this case shows the policies had the unintended effects on rural populations of adopting the policies to local needs and traditions. This development has far-reaching implications for the regime's current ability to implement population policies on a voluntary basis. Case 2 looks at the practice of PNT and decision-making about pregnancy in India. In a political context that discourages large families, and has poverty and limited healthcare access, Case 2 shows the complexities around the great concern families in India have about giving birth to healthy, male children, and the 'autonomy' counselling gives to women to decide about the outcome of their pregnancy after genetic testing. Case 3 concerns the question of abortion in Japan. In Japan, the Eugenic Protection Law was abolished in 1996 (coinciding with the implementation of the new eugenic policy in China). The law stipulated conditions to legally sterilise those with hereditary genetic disorders in order 'to prevent birth of inferior descendants from the eugenic point of view'.[2] Even though, until recently, the Japanese state supported the eugenics law, we witness that some women opt to give birth to fetuses tested to be positive for Down syndrome and other anomalies. Although this concerns a small minority compared with China and India, where cases of knowingly giving birth to affected children are rarely identified, it is significant. Case 3 explores why and how this reproductive decision is made. In short, the case studies will give insight into how life assemblages condition decision-making about reproducing populations.

Case 1: Reproductive Strategies in China: From New Eugenics to Lineage Enhancement[3]

In China, the historical background is of great importance to understand the coercive and large-scale nature of the shape population policies took on in the one-child policy, developed in the 1970s. The first comments on population issues in the upper echelons of the Chinese Communist Party date back to Deng Xiaoping's statement on the desirability of contraception in 1954, and relatively mild population policies were implemented in the 1970s in various regions (Greenhalgh, 2008). Large-scale population policies started to affect compulsory family planning only after 1979. Although the implementation of birth control policies varies by region, in general they promote the deferment of marriage and childbearing, fewer, but healthier, births and prevention of genetic defects. They advocate the practice of 'one couple one child' and encourage birth-spacing for those who may have more than one child (Hesketh and Xing, 2000). The one-child policy was criticised internationally and nationally for its infringement of human rights and the violation of women's reproductive choices. Fierce debate arose again when the New Eugenics Law, which made premarital check-ups mandatory, was ratified by the Chinese parliament in late October 1994. Although its name was altered to the Mother and Infant Healthcare Law (MIH Law), criticism continued after the policy became effective on 1 June 1995 (Dikötter, 1998). Premarital check-ups involved personal birth applications for both spouses, written comments from village committees, neighbourhood committees or work units, and written approval from township (town) governments (Scharping, 2003). At a local level, however, it was especially the examination's quality, its price and its implementation that were challenged. Local dissatisfaction and international criticism led to a revision of the Marital Registration Law in August 2003, changing the mandatory premarital check-up to a voluntary examination. Nevertheless, the importance of this check-up is still emphasised by the MIH Law, and the references that non-Chinese and Chinese scholars make to it as 'eugenic birth' reflects both official propaganda and the handbooks (Kipnis, 2007).

This case study draws on both a systematic reading of the handbooks and on interview data. In my exploration of eugenic birth and fetal education in the handbooks, I aim to show that these concepts belong to a state discourse. In the context of the exploration of new means of reproduction, such as IVF, the position of rural women and their spouses has to be understood in the light of the moral pioneering of family households. I randomly selected the 18 handbooks for eugenic birth and fetal education from one of Beijing's large bookstores (Wangfujing Bookstore), from the section reserved for hundreds of books on childbirth and fetal testing. In each of the selected books, I examined the standard sections dealing specifically with methods of eugenic birthing and fetal education. The books' authors are practitioners and medical scholars in the field of reproduction and family planning, and their opinions represent dominant state views of reproduction. While I lingered repeatedly in this section of the bookstore, I noticed an average of 20 young persons browsing there, some of them purchasing a copy of one handbook or another. In a mini-survey conducted in May 2009 among urban women from urban areas (mainly Beijing), 30 were asked if they consulted these books before or after marriage. Six of them replied in the affirmative, but not all had bought the books themselves. Others consult magazines on the same topic or talk with their friends about reproduction and the planning of childbirth. I also conducted a study of infertility in urban and rural areas, in which conversations were held with over 20 couples from rural areas and 30 from urban areas, some of whom met in IVF clinics, while others were introduced through friends and acquaintances. Conversations related to their motivation to go for IVF treatment, their experience with infertility and the choices available to them. Eight interviews with genetic counsellors-cum-infertility doctors were conducted, most of which pertained to problems faced by infertile couples in the countryside and led to interviews with IVF clients.

In 2001, the State Council issued a ten-year plan on children's development (2001–10), which was supposed to raise premarital check-up rates to 80% in urban areas and 50% in rural areas in 2010 (Anonymous, 2007). After the amendment of the Marital Registration Law in October 2003, however, premarital testing became voluntary, and the number of couples willing to undergo premarital check-ups dropped abruptly. While the percentage of would-be couples who had a premarital check-up before 2003 was estimated to be 68% nationwide, in 2004 less than 10% of them underwent premarital health checks, with the rate dropping to less than 3% in 2005 (Anonymous, 2005a). Birth defects rose steeply between 2002 and 2005 (Anonymous, 2007). The Beijing Public Health Bureau revealed that from 1996 to 2002 15,794 people were found during premarital check-ups to have hereditary diseases. Among them, 695 were diagnosed as 'not suitable to have children' (Cao, 2004). Worries about the increase in the rate of birth defects led a number of Chinese People's Political Consultative Conference National Committee members, including Zhao Shaohua, vice chair of the All-China Federation of Women, to propose amendments in the marriage registration regulations in a bid to make premarital health checks mandatory again. Wang Longde, Vice Minister of the Ministry of Health, announced that China might re-impose the premarital health check-up system (Anonymous, 2005b). The argument made was that had the check-up rate been higher, the birth of a great number of 'defective' children could have been prevented. The high rate of birth defects is widely attributed to people who are ignorant about genetic disease and do not understand the use of the tests. For this reason, the reinstatement of mandatory premarital check-ups has gained much support, especially in political and medical circles. Zhang Xizhi, a doctor with the Maternity and Children Care Centre of Hunan Province, explains: 'The policy of voluntary check-ups is humanistic, as it aims to respect the individual's privacy. However, due to a weak awareness among the public, especially in rural areas, people tend to ignore the significance of such health benefits' (Cao, 2004).

Such specialists further hoped that raising people's awareness of congenital defects and infectious diseases that can harm childbirth will persuade couples to have a premarital check-up. According to the handbook edited by Zhang Qingbin (2007: 31), the premarital check-up consists of:

A) Investigation of the couple's disease history: history of family disease and genetic background; history of innate diseases; date of birth; health habits (drinking, smoking and so on); menstrual history; previous marriages;

B) A general health check-up: checking of blood type compatibility; chromosomal and genetic diseases, creation of a genealogical chart;

C) A health check-up of the reproductive organs;

D) Laboratory tests;

E) Marriage advice.

Critics comment that the absence of genetic testing from this list means that such check-ups may give couples a false sense of security. Furthermore, many patients complain that the check-ups in some hospitals, especially rural ones, are a mere formality and a lucrative source of income, while diagnosticians in rural hospitals suggest that a lack of funding, reagents and testing equipment impedes effective technological control of the genetic make-up of Chinese citizens (interview Xu, Shanghai, 28 April 2007). In other words, even though many physicians and geneticists would support the state's eugenic policies, there is insufficient capacity for its implementation (Mao and Wertz, 1997; Mao, 1998; interview geneticist Tang, Beijing, 14 May 2007). In an attempt to increase the efficacy of the policy, some clinics have resorted to lowering the price for premarital check-ups or provide them for free (Cao, 2004; Anonymous 2007).

Handbooks for eugenic birth and fetal education, available in every standard bookshop in the country, aim to make young couples eugenically aware. They all point out that eugenics begins with partner choice. Apart from paying attention to character, age and socio-economic background, most handbooks emphasise the compatibility of the couple's health and their intelligence. The handbooks also advise on when to give birth and how to avoid children with mental or physical disease. Eugenic birth is directly associated with 'fetal education'. Fetal education is considered to be of increasing importance in a national and global situation of fierce social and economic competition. For this reason, the fetus should be trained from conception. The handbooks frequently refer to the age of 0 as the traditional beginning of life in China, which is why in China babies are one year old at birth. As a method, fetal education uses the stimulation of six senses (touch, smell, hearing, sight, taste, spinal sensation) using music, massage, petting, talking, reading, warmth, nutrition, chewing and breathing (Sleeboom-Faulkner, 2010a). But, owing to their focus on urban areas, they augment an already deep schism between urban and rural areas. In their advice and their argumentation, the handbooks presume China to be a modern welfare society with widespread use of advanced reproductive technologies; they presuppose its population to be keen on improving the quality of the Chinese stock; and assume that Chinese couples will increasingly resemble what are imagined as Western couples. In this way, the question of the impact of the handbooks on rural areas is largely ignored.

Interviews with infertility specialists and couples from rural areas indicate that severe problems exist in relation to access to healthcare, family pressure, son preference, poverty and the continuation of the family line. Yet, despite the focus on family planning, the handbooks do not mention how to deal with insufficient medical resources. Only one handbook makes special mention of home birth as an option for couples residing in the countryside (Xue, 2005: 45). It points out that medical facilities in the household may not be sufficient, 'although the familiarity of the home environment may be comforting' (Xue, 2005: 59). No mention is made of the financial conditions and healthcare facilities necessary when a premarital check-up reveals 'incompleteness' (*que bo shao tui* 缺膊少腿: literally, lacking an arm and missing a leg). Furthermore, there are no references made to disabled children born despite premarital or prenatal check-ups. The handbooks give the impression that the birth of any child with a handicap could have been prevented and is undesirable. This mode of presentation promotes the treatment of disabled children as second-rate citizens, and worsens the ostracism and discrimination faced by their families (Wu, 2007). The social isolation of families with handicapped members is especially problematic in the countryside. In one case, the birth of a child with Duchenne's muscular dystrophy isolated the entire family from the community, creating problems in finding marital partners for siblings and cousins, and exacerbating a heavy financial burden (Sui and Sleeboom-Faulkner, 2010b).

Most of the handbooks make distinctions between urban, rural and national minority areas. This distinction serves mainly to point out the backwardness of people living in rural and national minority regions.

The handbooks associate son preference with the countryside, and discourage it. They suggest that rural tradition treats only the male as family offspring, and call sons 'burning incense' (*xianghuo* 香火), for traditional custom prescribes that only a son can hold a memorial ceremony for the ancestors (Liu, 2005: 1–11). Despite the questionability of the idea that son preference is intensive only among rural women under the one-child policy (Wu and Walther, 2006), Zhang Jialin explains the forbidden practice of gender selection as associated with a socially and economically backward (*luohou* 落后) countryside:

> There are some male comrades who engage in 'favouring men and despising women' (*zhongnan qingnu* 重男轻女), a practice based on feudal thought, and especially widespread in the isolated countryside. When the wife gives birth to a baby girl, she gets cursed as a useless wife. If they cannot give birth to a baby boy to continue the family line, then she could even be divorced. Yet it is not possible to control the gender of the baby (2006: 44; transl. by Sleeboom-Faulkner).

It is also implied that people in the country do not know that women cannot determine the sex of the baby, in contrast with the urban female audience of the book. Yet, when it comes to modern medical insights, the handbooks provide do-it-yourself methods for gender selection as a solution in the case of X-linked chromosomal disorders (Lei, 2006: 18; Xue, 2006: 20–1). The handbooks especially associate isolated regions and national minorities with consanguineous marriage. Li Xingchun and Wang Liru are among the many editors of handbooks that relate the practice to the backwardness of rural areas and regard it as a sign of a lack of education (Qin, 2004; Tang, 2006: 25; Wu, 2006: 70; Xue, 2006: 10):

> Consanguineous parents have three times more chance of stillbirth and their offspring have a 150 times greater chance of having a genetic disorder. After consanguineous marriage became illegal in China, some people still do not understand the law; do not understand that consanguineous marriage would endanger their offspring. In backward regions and where old customs survive people think that 'adding family to family means continuing the family fortune' (*qin shang jia qin, bu duan qin yuan* 亲上加亲,不断亲缘; transl. by Sleeboom-Faulkner).

On the basis of China's old customs, people *presume incorrectly* that it is wrong for cousins of the same name to marry (*tang xiongmei* 堂兄妹), but that it is all right for cousins that do not carry the same name (*biao xiongmei* 表兄妹) to get married (Li and Wang, 2006: 29). Consanguineous marriage and gender selection clearly do not fit in with the idea of optimistic progress presented in the handbooks.

During the reform era, an important concern for rural households is the continuation of the patrilineal line upon which the household clan is built. In many places such continuity boosts the household's power base, status and finances. Thus, infertility is particularly problematic in the countryside, and sons are especially desired (Sui, 2007). When one of the partners is diagnosed as infertile, it is common to defer or cancel a marriage. The entire family household may be involved in the decision-making, with the aim of securing healthy male offspring. A doctor in a fertility clinic in Changsha told me about two cases of rural infertility, the first of which related to an infertile husband:

> In the countryside the situation is not very good. If the woman does not give birth to a child, then the husband will divorce her. Although a minority of highly cultured people do not care if they have no children, the great majority of the people are not so highly cultured... Many people have to come here [for treatment]. There was one couple from a country village that came here. The man was infertile. So I said, 'Well then, you need to have ICSI [intracytoplasmic sperm injection]'. When I saw the couple again, much later, I asked them, 'Why did you not come for ICSI?' The husband said that it was too expensive, and they could not afford it. At the time, he had a carpenter building a house for them; he made the carpenter sleep with his wife...But this child was handicapped. The child died. After that, they came to me again, now for ICSI. You can imagine how difficult that was. The husband just told the woman, 'Tonight you can sleep with him'. The woman has no say in it (interview Pu, Changsha; 28 March 2007; transl. by Sleeboom-Faulkner).

This case study tells us something about the views of a genetic counsellor on life in rural households. The genetic counsellor regards rural individuals as uneducated and backward, giving no thought to the political and financial pressures under which the household is placed to continue the family line. Members of the family household are denied any rational agency.

The second case study concerns an infertile wife:

> Her mother-in-law and husband consulted each other, and asked a peasant girl to live with the husband. She gave birth to a child.

Now the husband treated the girl with more intimacy than he did his wife. But then the girl wanted the husband to get a divorce or give her a lot of money. These kinds of circumstances still exist widely in China (interview Pu, Changsha; 28 March 2007; transl. by Sleeboom-Faulkner).

In this case study, members of the patrilineal household—the mother-in-law and the husband—decided to create an offspring by engaging a rural woman when the wife was assumed to be infertile. The feelings of the wife counted for little, but were not ignored, as the woman hoped to raise the child as her own. The rural woman became demanding and did not plan to give up her child. The husband then decided to stay with his wife and try IVF, as he chose not to affront his wife's family. There is no simple formula regulating rural reproductive life, divorce or abortion; family and household situations are complex. Decision-making in rural areas is just as rational as it is in cities. But the tendency in rural areas is to make decisions in favour of the continuation and sustenance of what is regarded as the good of the family household.

When entering the marriage market, a young woman needs to be healthy and fertile. If she decides to go for a premarital check-up, she runs the risk of getting herself and her family devalued as a low-quality human commodity (Sui and Sleeboom-Faulkner, 2010a).

The example of Hu Xin illustrates this. This young woman never intended to go for a premarital medical check-up, for she knew that what could be revealed 'could drive Mr Right away'. But a health check after her marriage revealed that she carried the hepatitis B virus, as do about 10% of the Chinese population, according to the Ministry of Health. Hepatitis B can lead to liver cancer, but inoculation is cheap. Nevertheless, many people think that the virus can spread through casual contact and this has led to widespread discrimination against hepatitis B carriers. Hu Xin's husband instantly filed for divorce, as he was disappointed in Hu's apparent concealment of the truth (Wu, 2007). For mental disorders, similar circumstances arise. One of the handbooks warns that 'if your family has a history of schizophrenia or manic depression, you have a higher than average risk of producing affected offspring' (Wu, 2006: 158). If kept secret, you have a good chance of finding a spouse and still having children, but, if revealed, it warns, you may not succeed in finding a partner. Worse, you may have unhealthy children. In short, being aware of eugenic knowledge, no matter how imprecise, unscientific or tendentious, has consequences for the marital prospects of members of rural family households. For those who depend on the family household for their material and emotional survival, getting a premarital check-up can potentially reveal damaging information about the family lineage.

Eugenic ideas about family planning in China are intimately linked with the organised control of human birth, ideas about improving the quality of the Chinese population, and a staunch belief in enlightenment ideas of progress and rationality. China's policy is based on popular views of social evolution, instrumental views of birth and the assumed capacity to raise the quality of the population. Nevertheless, the use of compulsion in birth planning is mostly rejected in favour of persuasion and emancipation by means of public education and propaganda, so that voluntary decision-making and responsible individual choice are important moral considerations under this socialist regime.

In rural areas, however, family households prioritise their own ways of continuing the family lineage's interests. Rather than leading to a high demand for premarital check-ups, family planning propaganda has tended to encourage prospective couples to make strategic and rational choices to maximise their own and their family household's interests in negotiating marriage. The notion of lineage enhancement contrasts with notions of laissez-faire eugenics or flexible eugenics, which emphasise factors underlying individual choice in reproductive decision-making. The notion also contrasts with concepts underpinning state eugenics policies that are based on a discourse of nation-state building and the enhancement of the national and racial Chinese stock. Instead, lineage enhancement implies that reproductive choices are partly shaped by state eugenic policies but are, at the same time, a historical product of Marxist state ideologies, cultural traditions and local relations of power, including that of male domination.

Case 2: Private Eugenics in India?[4]
(by Jyotsna Agnihotri Gupta)

Population policies, poverty and healthcare access are major factors shaping the choices individuals and families have in which to give birth to children. Of India's population of over 1.2 billion, 70% live in rural areas where basic healthcare services are unavailable, inadequate or inequitably distributed, with class and literacy/education being important factors determining access to medical settings. In 2009, the total number of registered births was 22,500,000. Perinatal causes, respiratory infections, diarrhoea, measles and malaria, infectious diseases and, above all, under- and malnutrition are factors responsible for the high infant and child mortality rate. The neo-natal mortality rate (less than 29 days after birth) was 34 per 1000 births; the infant mortality rate was 50 per 1000 births; and the under-five mortality rate was 64 per 1000 births (Sample Registration System 2011 Registrar General of India). As more children are surviving as a result of public health measures, including immunisation programmes, geneticists claim that India, like many other low- and middle-income countries, is undergoing an epidemiological transition, with a relative increased role of genetic factors in health and disease (Verma and Bijarnia, 2002: 192–6). Primary healthcare advocates, however, argue that a major share of perinatal morbidity and mortality, as well as infant and child mortality, may be ascribed to under- and malnutrition of both mother and child, and a lack of simple ante- and postnatal services within primary healthcare.

India has a large, perhaps the largest in the world, unregulated, poor quality, expensive and predominantly private health sector, and an inadequately resourced, selectively focused and declining public health sector. Super-specialty services such as genetic testing are left largely to the private sector in order to reduce government expenditure. Clinical facilities for genetic investigation in India are rather scanty and expensive, with a shortage of trained clinical geneticists. Owing to a lack of standardisation of laboratories, test results are often not accurate. Since 1952, population reduction has been the main objective of population policies as the government identifies 'population explosion' as a major obstacle to economic development; therefore, most health funding has been used for this aspect of family planning. As the number of children per couple is going down (currently approximately 2.2), especially in urban middle- and upper-income households, there is more emphasis on 'quality' rather than 'quantity'. PNT plays upon two fundamental desires of all couples: to become parents and to produce healthy and 'normal' children (Gupta, 2007).

Public awareness regarding the role of genes in the incidence of disease and the possibility of making use of PNT is increasing; it is associated with modernity and good parenthood. India has never had an explicit eugenic policy, although 'eugenic' ideas are widely prevalent among lay people, particularly the elite, as well as medical specialists and population planners. The risk of giving birth to an affected child is formally a valid reason for abortion, legal up to 20 weeks under the Medical Termination of Pregnancy Act of 1971 and the Amended Act of 2002. Owing to the publicity and proliferation of PNT for sex selection (illegal under The Prenatal Diagnostic Techniques Regulation and Prevention of Misuse Act of 1995 and the Amended Act of 2002), there is widespread awareness regarding tests such as ultrasound and amniocentesis (known in common parlance as the 'boy/girl test') among a large segment of the population. This case study examines factors important in reproductive decision-making among high-risk pregnancies. The study is based on a literature study and fieldwork at a large private trust hospital in New Delhi in two phases (January–February 2006 and January–February 2007), permission for which was obtained from the ethical committee of the hospital. As there are only 25 genetic testing and counselling centres in India, a high number of women receive services here. The fieldwork comprised observation of client–provider interaction during genetic counselling sessions and semi-structured interviews using guiding questions. The majority of women came for genetic counselling after an indication of higher risk identified through the triple marker test, advised by their gynaecologist. Others had experienced primary amenorrhoea, repeated miscarriages or were 'high-risk pregnancies' owing to advanced maternal age (the cut-off point used is 35 years), a family history of certain diseases, such as identifiable syndromes, mental retardation and so on, or those who had a previous affected child.

Much has been written about the overt and covert pressures on Indian women from family and society regarding the number and sex of their children; little is known about the pressures on women to give birth to a healthy child. Not only do families face the financial burden of caring for an affected child, they also meet with the stigma attached to giving birth to a child with a disabling condition. Religious and cultural values ascribed to 'normal' and 'disabled' children prevalent in society play an important role. One interviewee, named Jaya, who worked at the hospital's reception desk, had a five-year-old daughter with a clubfoot. Pregnant with her second child when interviewed she was very eager to relate her experience:

My first child was born three years after marriage. There was family pressure on us to produce a child. My in-laws said, 'You have produced this baby after three years!' It hurts when people talk like this about my child. I had serious depression after my delivery. My family, especially my mother, consoled me. I asked my gynaecologist why she had not been able to detect it with ultrasound. I felt very sad. What will I tell the world? I wonder what kind of deeds [karma] we had done to deserve this. I would feel that my child was the worst. I felt very down, although as a couple we're the best in the family. We're both so intelligent, too. I was very depressed. Suppose there's such a problem again; that's why we don't want to take the risk (interview Jaya; 31 January 2007; transl. by Gupta).

Jaya informed me a couple of days later that she had aborted the pregnancy. What was remarkable was that all the interviewed women (and their husbands if present) said that in case of a 'positive' test result they would abort. It would be a very painful, but pragmatic, decision based on concern for the child's future and their own. Only a few postpone the decision:

We will do all the tests, so as not to take any chance. We will abort if there is any problem (interview Ashok, husband, undergraduate, businessman; 12 January 2007).

The standard answer indicated that the couple was not prepared to take any chance:

The doctor says it is only a 2% risk, which, too, we want to avoid. I'm not being emotional, but practical. It is the height of stupidity to knowingly bring a disabled child into the world. We don't want the child cursing us. I'm a teacher who has come across Down's syndrome children. They are slow learners. I have one in my class…I come across thousands of kids at school with different problems. We've thought about it. [In case the fetus is affected:] Why bring a child like this into the world? In case it is born, everyone will do everything for the child. Also, money is not the issue. But, after us [our death] who will look after the child?' (interview Seema, schoolteacher; 12 January 2007).

I have discussed it with family members; not only my husband and I, but the whole family feels I have this opportunity for testing. The decision thereafter will be based on the degree of severity of the condition. It will save me and the child from trauma (interview Renu, postgraduate, employed; 22 February 2006).

Women feel burdened by the responsibility for decision-making, torn between what is expected of them by others, including the doctor and their family members, and their own emotional response to having to terminate a desired pregnancy:

If the test had been done earlier it would have been better; now, I can hear the heartbeat. There is also an emotional attachment (interview Seema, schoolteacher; 12 January 2007).

I want all the works, tests to rule out Down's syndrome. My risk due to maternal age is 1:12; the test shows 1:320. You decide for me doctor, you're the expert...I know that children with Down's syndrome are very happy, but I think I'm old, and I think it is immoral to bring disabled children into the world as an older mother (interview Rita, businesswoman, during genetic counsellor–client observation; 29 January 2007).

Dr Amar, a geneticist and genetic counsellor at a private hospital, discerns a difference in reactions when confronted with the birth of a disabled child between families in India and in the West:

Most people know some family or other with a disabled child and the care burden it brings...There is no rejection of affected children once born; there is more acceptance here than in the West. There is hardly any institutionalisation here; the child is taken care of at home... Unlike the West, if there is an abnormality in a child the whole family comes for counselling. It is very rare that only the couple comes; in fact, the couple is the quietest. We try to take away the guilt of having given birth to an abnormal baby (interview Amar; 3 March 2005).

His colleague, also a genetic counsellor (Dr Anjali), added:

The whole emphasis here is on preventing the birth of an affected child, because of financial reasons...Also, there is a lot of social stigma attached to giving birth to such children, both from the mother's and in-laws' side (interview Anjali; 24 February 2005).

As is clear from the quotations above, both pregnant women and genetic counsellors are aware of the social stigma attached to giving birth to disabled children and the blame apportioned to mothers by the family, especially the marital family.

An important factor in the uptake of tests is the cost of tests and clients' capacity to pay, which is a major obstacle for some. Most women mentioned that the cost of tests is prohibitive; only their/husband's employers reimburse a few for costs incurred. As one woman said: 'Screening tests are not affordable but necessary, whether refundable or not' (Aruna, student of naturopathy; interview 27 January 2007). Several mentioned that their family had pitched in by selling family land or assets, and giving or loaning them money. The high cost of medical care is one of the leading causes of indebtedness in India (Gangolli et al., 2005).

Clinical encounters are characterised by power hierarchies, based on a perceived social distance by clients between themselves and the service provider, mainly due to the medical authority of the latter and their own inability to understand medical jargon. Clients with high education seem to gain more information and satisfaction from counselling sessions than do those without. Clinicians tend to give more time to the latter in explaining and answering questions, and yet they seem less satisfied. Dr Rajni, genetic counsellor, said:

> We generally give non-directive counselling. Depending on the understanding of the client, one has to change the counselling, but we remain within these parameters. Much of the work consists of providing information—oral and written—and helping clients to come to a decision. What is the risk of conceiving [another] affected child? We give them a written page with advice containing the main summary of recommendations. In some cases we give them leaflets on certain diseases. Many get information these days from the Internet; other sources are their own obstetrician and family doctor. Pregnant women carrying a fetus diagnosed with an abnormality come for confirmation and the course of action to follow. Some are prepared, and very apprehensive. Patients wonder 'Why has it happened to us?' Some put all their faith in us and listen to us; some check and recheck; doubting everything we say. Some are angry at the doctor about what has happened. Patients are more worried than need be; we help to relieve their tension (interview Amar; 24 February 2005).

In an attempt to make sense of how to act on the basis of what was articulated as unexpected risk knowledge, many couples reacted—almost by reflex—by asking the health professional what they would do in that situation:

> The only variation with the West is that patients insist on directive counselling. 'What would you do, Doctor Sahib, if you were in my place?' We tell them that other couples in their situation would abort. We tell them the risk; we don't tell them to abort (interview Dr Amar, genetic counsellor; 3 March 2005).

The genetic counsellor's advice is most important, but the decision apparently rests with the family:

> Parents want more directive advice. Non-directive advice is not acceptable, because they are not so literate about science. They say they will

go and discuss with the family whether to go for an amniocentesis or CVS [chorionic villus sampling]. Family members' views are very important. They discuss with their own parents. It remains between the husband, wife, and parents; they generally don't want to let other family members know if the diagnosis is positive (interview Dr Amar, genetic counsellor; 3 March 2005).

The pressure on women both to undergo tests and to opt for abortion in case an anomaly is detected is high. Most pregnant women stated that the decision regarding testing and the course thereafter would be taken on the advice of the genetic counsellor in consultation with their husband and other family members. Some of the answers to the question 'Who will decide whether the pregnancy would be continued?' were as follows:

> For most of our patients it is a 'family consent'. The whole family is involved and comes along (inteview Dr Rajni; 10 January 2007).

> The decision regarding testing depends upon the doctor and whether we can afford it. Later, it also depends upon the test result (interview Geeta, MBA, university lecturer; 24 January 2007).

> I'm not scared, but if there is a problem I will act according to what the doctor instructs (interview Aruna, student of naturopathy; 27 January 2007).

Few women have knowledge of testing procedures beforehand. Most respondents, including the fairly educated ones, complain about the lack of adequate information provided and empathy by the service provider:

> The information given is so last minute…You have to be so tough to get information whereas it should be publicly available, so that you're not anxious (interview Vibha, postgraduate, resident in the USA; 7 February 2007).

Institutional facilities for care of the disabled are very few and inadequate, and parents of children affected by genetic diseases get little financial support to look after them:

> In this country they do not get money from the government to look after an affected child. They express their difficulty in looking after an affected child. Having a child with a genetic disease is quite tough. The government is increasing positive measures for the disabled;

there is a 3% reservation in jobs for them. For mentally retarded children the government provides free education. The family gets a tax rebate if it has a child with an incurable disease. But the social services are inadequate for such a large country and few services are good (interview Dr Amar, genetic counsellor; 3 March 2005).

The perspective of disabled people is often missing in the information given to women/couples seeking genetic counselling. According to Juhi, university lecturer, disability rights researcher and activist, and physically challenged:

> It seems somewhere counselling has failed...The professionals don't give a realistic picture. They paint lives with disability as very black. If disabled people would be part of a counselling team people would think differently; disabled people should be included in it...(interview Juhi; 1 February 2007).

In Juhi's view, more attention needs to be paid to tackling social and environmental factors that cause disability and make living difficult for people with impairment.

Women in India neither have sufficient information to give informed consent nor the autonomy to exercise it. Case 2 shows that Indian families decide on whether to have an abortion. On receiving test results indicating the slightest risk of fetal disability, families usually decide that the woman should abort. Although genetic counselling gives women the responsibility to decide, they feel burdened by the responsibility for decision-making, torn between what is expected of them by others, including the doctor and their family members, and their own affective response to having to terminate a desired pregnancy. Of principal importance is that positive test results remain within the family. Considerations of the moral status of the embryo seem to be irrelevant in the decision-making, although unintended disabled children born are taken care of in the household.

Case 3: Birthing Decisions and Down Syndrome in Japan[5]
(with Masae Kato)

Japan is one of the few countries in which official laws on eugenics have been passed in the form of the National Eugenic Law (1940)[6] and the Eugenic Protection Law (1948) (Ministry of Health and Welfare 1940, 1948).[7] These laws articulated legal procedures for the sterilisation of disabled people with the intention of enhancing the quality of the nation. As a consequence, there has been active public debate about the practice of selective abortion for many years. Discrimination against disabilities (shôgai) is also deep-rooted in Japanese society. At the same time, reproductive genetic technology is developing rapidly. The possibility of diagnosing physical defects in embryos through PGD is making it possible to avoid abortions, and allows both doctors and individuals to select a 'high-quality' embryo rather than depend on selective abortion. However, when it comes to actual decision-making in the course of pregnancy, not all individuals choose to use testing methods, such as amniocentesis, or to terminate pregnancy, even when prenatal diagnosis results are positive for a problematic condition. In the decision-making process they struggle, weighing a number of factors against one another, and, finally, some decide to terminate the pregnancy, while others decide to continue it. Even though some decide to terminate the pregnancy because of an unusual adverse fetal health condition, such decisions are not without agony, a sense of guilt, apology and attempts at self-justification as individuals afterwards attempt to regain balance in their lives.

This case study discusses socio-cultural factors in Japan that do or do not lead individuals to terminate pregnancy when an anomaly is found in a fetus. The decision to continue with a pregnancy after a positive test result is not usual. There are no official figures on selective abortion, but the 11 obstetricians interviewed stated that 80–90% of women decide to terminate pregnancy in cases of an anomaly of the fetus. The numbers of those deciding not to terminate pregnancy may look small but are significant if compared with other Asian countries, where we have not been able to find such cases. An understanding of the factors that bring about the decision not to abort may reveal something about alternative attitudes to childrearing. The case study is based on a literature study and fieldwork, including five in-depth interviews held in early 2006 with women who, after a positive diagnosis of Down syndrome, decided not to terminate the pregnancy. It also draws on 46 background interviews with women/couples with experiences of prenatal diagnosis, pregnancy and parenthood, and 11 obstetricians in Tokyo and Osaka. Consent was obtained to interview and use the anonymised transcription of the data—pseudonyms replace the names of interviewees.

This case study explores, why if women have the option to abort an affected fetus, decide to give birth. Japan is not known as a country with religious dogmas concerning the fetus, and abortion is still in the penal code. So why would couples opt to continue the pregnancy? Among the relevant factors, reproductive history was a main one in women's decision to give birth. Reproductive histories include a history of infertility treatment, stillbirth and already having a child with a disability. Thus, Chie, who had received eight years of infertility treatment and who found a heart deformity in her fetus during her first pregnancy was inclined to give birth also to a disabled child. Family structure also plays a role. In the case of Akemi, who had experienced a stillbirth in her second pregnancy after having given birth to one son, she decided to give birth to a child with Down syndrome (her third pregnancy). She worried that she would deny the value of her son's existence if she gave up the pregnancy. In addition, she had always wanted to provide a sibling for her son, but hitherto she could not because of infertility problems. In contrast, Hideko, who already had one child with Down syndrome, asked 'How can I show my face to society if I have another child with Down syndrome?' Here, reproductive history and family structure were decisive factors in having amniocentesis and terminating the pregnancy.

The space between the child born with an anomaly and the following pregnancy influenced decision-making. In the beginning, parents would often be surprised by the birth of the child with an anomaly but, with the passage of time, they started to feel more confident about dealing with the child. In these circumstances, some of them chose not to have prenatal diagnosis, thinking that they would be able to accept any child.[8] The timing of finding an anomaly in the course of pregnancy plays another critical role. Chie found a heart disorder in the fetus by means of ultrasound at 22 weeks of pregnancy, which was just too late to have an abortion according to the law. Her obstetrician said, however, that she could find a doctor to do an abortion if she wanted because the disorder was severe. But she decided to keep the child as she could not bring herself to end its life. The desire to have a child had made her and others decide to continue with the pregnancy, but had the discovery of the disorder been earlier, or had she not felt the fetus move, the life would have been terminated (see also Ozasa, 2006).

Among others of influence in the decision-making of the woman, it was the husband who was most important, and all reached agreement. Three out of the five interviewees said that their husband saying 'Giving birth is fine (*unde mo ii yo*)' had been the green light for them to continue with the pregnancy. Akemi found out during her pregnancy that her fetus had Down syndrome. She decided to give birth, but her parents-in-law and her husband opposed this decision, which made her hesitant:

I angrily told my husband that I would divorce him and give birth to my child. He immediately apologised and supported my decision.

I really had threatened to divorce him then. My determination was that firm (interview Akemi; 8 May 2006, Hiroshima; transl. by Kato).

Akemi described herself as a determined person, but still she explained that her husband's support was essential encouragement to stick to her decision, despite opposition from her husband's family. She continued:

> My husband's parents opposed my decision. But I gave birth. Then, when the child was born, they started to intervene in my plans on how to raise my daughter. I do not talk with them anymore; I do not let them intervene...I must say, however, that I could always remain strong because of my husband's support (ibid).

Akemi's husband supported her decision, which encouraged her to continue with the pregnancy. There were instances among the other 46 cases in which parents or parents-in-law played a dominant role in the reproductive decision-making of young couples, not only about the desired health condition of the prospective children, but also on the ideal number of children or spacing between the birth of the children.

Obstetricians also influenced the decision-making of women and the formulation of opinions by husbands. Hideko and Emi, who learned of chromosomal disorders in their fetus, were told by their obstetricians that 'in many cases like yours, people decide to have an abortion'. Hideko did not choose to have an abortion, but she was angry at what she heard, describing this attitude as 'ignorance of the value of life' and too 'guiding'. Emi also wanted to continue her pregnancy because she had become pregnant after ten years of infertility problems, but her obstetrician told her that 'this type of chromosomal disorder is more serious than normal ones, leaving the child unable to speak or even stand up'. There were no means to test the accuracy of this diagnosis, but it was possibly correct. This statement, Emi explained, was the decisive factor in making her final decision, although she was not entirely behind it. Discussing the doctor's view with her husband, she decided to terminate the pregnancy. Her husband said, 'The obstetrician knows most about clinical matters and they told my wife to abort'. Then, 'I had no other way but follow the advice'. However, in the cases of Akemi and Chie, who gave birth to children with anomalies, it was their obstetricians who told both of them that giving birth to the child would give them confidence and encouraged them to continue pregnancy. In this way, it is possible to see how specialists' words may be influential in formulating the opinions of both husbands and wives.

Emi's greatest concern in deciding whether to continue the pregnancy was about childrearing. As her obstetrician told her that her child would probably not be able to stand up or speak, Emi worried that being a full-time caretaker of her disabled child would drive her mad. The other four interviewees expressed concerns about childrearing, but they were more occupied with the anxiety of whether or not the fetus might survive with the identified disorders. A closer look at the interaction between a husband and wife reveals how gender roles play a role in the decision-making process. Not only these five interviewees, but other interviewees, too, generally took it for granted that it is women who take care of the children. During the interviews, all the husbands mentioned that it was their wives who should make the final decision of whether to have a pre-natal diagnosis and the decisions thereafter. All of them indicated that they would support whatever decisions their wives made. Interestingly and contradictorily, however, women stated that their husbands had the greatest influence on their decision-making. Among the other 46 individuals, too, when husbands were asked why they thought their wives should make the final decision, a range of views emerged. 'It would be my wife who would have to look after the child if it was disabled', said Tatsuo, the husband of Tatsuko. 'It would be my wife who would have to give up her work', said Fumio, the partner of Fumiko. 'It is women who create new life', 'we men do not know the pain of amniocentesis, abortion or delivery' said Hiroshi, husband of Hiroko. We can see here the effects of a strict gender division of labour. Such attitudes were prevalent not only among husbands but wives also.

The five women interviewed did not have careers as such, and none of them was the major wage earner in the household. Nevertheless, even among the 34 of the 46 women surveyed who had a career, it was almost always the woman who was prepared to reduce her working hours. Some of the women felt they had let their husbands down by having a child with a birth defect. One woman in the larger sample explained 'I was so sorry not to be able to let my husband have a normal child'. Emi expressed 'gratitude' to her husband, saying: 'Many men would blame their wives for the abnormal condition of the fetus. But I am so grateful that he did not do this to me'. Women's gratitude to their husbands also contained economic factors. As Tsukamoto (2005) notes in her research, it is usually husbands who pay the financial costs of the pregnancy, delivery and rearing of the child. In this situation, women may lack confidence and feel dependent upon their partners. Together with the traditional ideal image of women being fertile and bearing healthy children, this triggers the attitudes of women seen in the interview cases.

Analysing the factors influencing reproductive decisions among women with fetuses diagnosed to be positive for Down syndrome in Japan, we find that there is a notion of pregnancy as 'parenting' among women and also among medical doctors. This may derive from cultural notions that see children as choosing to be conceived, rather than being just the choice of the parents. Among couples in the larger sample who chose 'good-quality' fertilised eggs after IVF, many kept thinking nostalgically of the frozen fertilised eggs that did not develop or that were not chosen, wondering if they were coldly ignoring the will of the fetus and if they were good parents leaving embryos in that manner. They call the fertilised egg 'my child' (*watashi no ko*) or 'our child' (*watashitachi no ko*), too.

Japanese society places a high premium on women as social agents producing high-quality human capital in the form of well brought up and educated Japanese children. In Japan, the term *sengyô shufu*, or 'professional housewife', describes women who do not work outside the home, but devote all their energy to supporting their husband's career and their children's educational, emotional and physical advancement (Ivry, 2007), with the mother and child being seen as one entity rather than two separate beings (Ehara, 1985; Ivry, 2007). This socially expected role of women extends to pregnancy, too. During pregnancy, some half of the 46 interviewed women withdrew from society during pregnancy, making it a habit to talk to the fetus or otherwise engage with it, and paying close attention to diet and other physical factors. Obstetricians in Japan frequently recommend that pregnant women refrain from work as much as possible during pregnancy (Ivry, 2007).

It is clear, that in Japan there is the complex relation between PNT and issues of eugenics. Japan is one of the few countries that have had official laws on eugenics and, consequently, there have been active debates about whether PNT constitutes eugenic practice (Ehara, 1985; Tateiwa, 1997). To date, however, there exists no consensus on whether prenatal tests, including amniocentesis, constitute 'eugenics'. Many obstetricians are frightened of being labelled 'eugenic practitioners' by the media (Kato, 2007). Moreover, in Japan, abortion is legal only in cases where the pregnancy is the result of sexual assault or when the economic position of the would-be parents is so grave that it would endanger the life of the child. Strictly speaking, therefore, selective abortion is not legal under current abortion law. When it is practised, it is done under the loophole of the economic reasons clause. Prenatal tests and selective abortion are therefore practised on a fragile basis in Japan, and the responses of parents should be understood in this context.

The main factors for couples to decide not to abort an affected fetus were their reproductive history, fertility, family structure, timing and spacing. But it is especially when the mother believes there is a possibility for the child to have a good life that the decision is made to give birth. Another factor is that the women that had already given birth to a child with Down syndrome believed that by aborting the child she would deny its existential value. This issue in Japan has attracted much attention in the light of the eugenic law, which had been heavily criticised by feminist and disabled rights groups, and was abolished in 1996. In Japan, the support of the husband is of great importance to the woman's decision, but not always crucial, and family pressure, though of great relevance, is not usually decisive in whether the pregnancy is completed. As the husbands generally regard childrearing as the domain of the wife, especially if they do not have a career, the wife usually plays the main role in the decision-making. But the women are often keen to consult and please their husbands.

Conclusion

Using the concept of eugenics in describing strategies of reproductive decision-making in Asian countries is misleading. The state eugenics prevalent in Germany and other nations in the first half of the twentieth century is hardly comparable with the Japanese eugenics law and the 'new eugenics' that have been relevant to policy-making in the second half of the twentieth century, as these nation-states do not aim for racial purification and are not politically motivated to commit genocide. The cases discussed above, then, focus on other forms of intervention with birth planning: the levels at which it takes place ranging from the state and the village community, to the household and the individual, and the influence of which ranges from the exertion of coercion to encouragement.

Case 1 discussed state population policy in PRC and the nature of its changing force and influence. Compulsory premarital testing from 1993 until 2002 and the one-child policy meant that the policy was carried out without the consent of the prospective parents concerned. The policy for most family households in China defined the narrow parameters between which families were expected to raise their children. But, over time, the policy was strategically adapted to the interests of individuals and household. The premarital testing information garnered from health examination was used to evaluate the quality of potential marriage partners and the quality of their offspring. Rather than leading to eugenic enhancement of the national Chinese stock, the life assemblage shows concerted efforts on both the level of the individual and the family to enhance the household lineage. In this context, many young couples became averse to testing as an outcome indicating disease could affect the fate of the entire family household. Owing to this fear, new public health policies involving genetic testing were hard to sell to the population, especially after compulsory premarital testing was abolished. Thus, policies against sex selection and appeals to undergo voluntary testing for thalassaemia and other anomalies gained very little response. Although a lack of information on such tests is one reason for this, it is also clear that families and individuals are worried about, and aware of, the potential misuse of test results. This trend may interfere with the intended effects of coercive state intervention with reproductive strategies of families.

Although, compared to China, India's coercive population policies have been less prolonged and radical, Indian families share with Chinese families (especially in rural regions) the relatively great say of family members, rather than the pregnant mother or couple, in reproductive decisions. Case 2 showed the great importance placed on genetic testing in India to avoid the birth of disabled children as a family affair. It seems that economic factors, lack of healthcare provisions and the social stigma of disability play a crucial role. Japan, a wealthy welfare society with near-universal healthcare access provides families and women with more leeway for giving birth to a disabled child. Case 3 illustrated that even after the eugenic law was abolished in Japan, women are still sensitive to measures that violate the 'human rights' of children and their karma, and sometimes decide to give birth to them. But it is especially couples that have experienced fertility problems that decide to bring the pregnancy to full-term.

In all three countries the family plays an important role in reproductive decision-making. And although formally women are supposed to have the final say in the decision of whether to give birth, in practice husbands and families are consulted, and play a main role. In China and India, however, the family plays a much more decisive role than in Japan. In Japan, women are generally aware of the meaning of directive counselling, and they know that the obstetrician's role is to give guidance, and not to interfere or direct the decision. In China and India, directive counselling seems to be more generally desired. But we don't really know to what extent this is the case from the point of view of women, as their position and wishes cannot be fathomed separately from those of her family. Women's reproductive pragmatism (Lock and

Kaufert, 1998) exhibits a flexibility rendering a delineation of her wishes from those of others impossible.

When confronted with a positive test of what is perceived to be a disablement, the majority of women in the cases discussed terminate the pregnancy. But when a disabled child is born, despite having taken precautions, family households tend to take care of it all the same. Stigma of disability is strong in China and India, to such an extent that it affects marriage prospects, social standing and the functioning of family members. In Japan, although stigma is also strong, institutionalisation and a greater awareness of disability have led to a diversification of attitude, varying from strong support to condemnation. Apart from the materially and financially privileged environment, the awareness of the value of life as appreciated by the disabled has contributed to the decision of couples in Japan to give birth to children with a disability. In all three countries, however, abortion of both disabled and healthy children leads to suffering of the woman carrying the child, especially when done under pressure.

Although the term eugenics cannot adequately describe both practices of coercive abortion and 'autonomous' choice at the same time, this does not mean that the pressures leading to undesired termination of the pregnancy and its consequences for the disabled population are in any way acceptable to them. Whether in the name of eugenics and genocide, or in the name of population quality and normal health, the existential tie between potential parent and child and the existential value of disabled persons are at stake all the same. As such, the three case studies do not only reflect various levels of decision-making, but also different existential values that have diverging implications for the life assemblages in the three respective countries.

The case studies exemplified life assemblages described on a nation-state level, primarily because the state has far-reaching decision-making powers where it concerns family reproduction and child-raising facilities. Case 1 sanctifies and proscribes fetal termination at state-level for the purpose of the improvement in quality of the collective population, but which at the family-household level is influenced by gender preferences and at the level of couplehood through strategies of protecting the reputation of individuals and families; Case 2 depicted a life assemblage in which families generally play a major role in making decisions concerning the interruption of pregnancy according to gender, health and economy; and Case 3 showed a life assemblage that regards the individual couple as having the last say in the decision-making struggle over the termination of fetal life on the basis of health, economy and values of respect for life. Although all three countries have come to regard reproductive technologies as means conducive to birth control, expressions of discomfort with such intervention as 'unnatural' were most salient in Japan. And although ideas about population control through forced birth intervention at a state level was strongest in China, both India and Japan have had to deal with similar notions in the past. The family has much reproductive power in the life assemblages in India and China, and, to a much lesser extent, in Japan, where couples can afford to take greater distance from the views of parents and in-laws. Although such comparison may provide us with an interesting list of similarities and differences, it is primarily useful as a resource for asking new questions about the introduction of reproductive technologies. These findings should be understood primarily in the context of the influence of relevant life assemblages.

3
Biopower and Life Assemblages: Genetic Carrier Testing in India, China and Japan

This chapter is concerned with the ways in which genetic knowledge in various Asian societies is part of varying life assemblages, conditioned by specific constellations of power. In Chapter 2 we saw how in life assemblages in China, India and Japan, apart from state-level pressures, community- and family-based interferences may be of great influence on the reproductive decision-making of women and couples. This chapter observes how these lower-level pressures affect practices of genetic screening, carrier testing and genetic testing, which also have implications for reproductive capacity and decision-making. The chapter further examines the question of how concepts of genetic identity are created through Asian life assemblages, and how these relate to the notion of biopower, which was developed largely in a European context of 'biobureaucratic institutions'.

Michel Foucault's work on the politics of the body saw the category of the 'body' as having expanding existential, bureaucratic, symbolic and politico-economic significance (Kohrman, 2005; Rose, 2007; Sunder Rajan, 2007). Foucault described the emergence in Western Europe of biobureaucratic institutions (clinic, psychiatric hospitals, welfare homes), and what he calls normalising discourses, producing modern apprehensions related to biomedicine, public health, demography and psychiatry. Foucault insists that power should not be described in negative terms (repression, masking, concealing), but in positive terms of the production of reality, domains of objects and rituals of truth. In the hegemonic order of things, dominant discourses render unthinkable the inconsistent 'truths'. For this reason, 'misrepresentation' and 'false consciousness' have no role in his analysis of this production; they are mere truth effects of power.

'Biopower' in Foucault's view promotes the signification and identification of differences among populations and between individuals.

The management of 'abnormal', 'unhealthy' bodies and the configuration of what is life, constitute defining instruments in the production of efficiency, progress and economic expansion (Kohrman, 2005: 6–8). Populations and individuals are reproduced together. At the level of the reproduction of populations, medicalisation is neither personal nor medical, but generalised. Thus, women are encouraged to produce healthy families with no medical liabilities. In the context of the reproduction of the family, women subject themselves to medical technologies, being disciplined into an ideology (truth effect) in which reproduction is considered 'natural' and fulfilling.

Foucault's insights into the development of biobureaucratic institutions, the normalising power of discourses and the productive nature of hegemonic power, although regarded as valuable, have been criticised with regard to the concepts of power, the state, ideology and agency. First, 'power' in Foucault's work is imbued with transcendental qualities. Foucault was so interested in demonstrating that authority was always relational and diffuse that he overplayed the state's insignificance as a sphere of socio-political formation (Agamben, 1998). Steven Sangren (1995) described how in Foucauldian cosmology power is de-subjectified, occupying the position of transcendence. In Foucault's essay, 'Governmentality' (1991[1978]), the state is no more than a constructed reality and an abstraction. Although it can be argued that understanding state power relative to other powers is important in a European context, this chapter focuses on the importance of distinguishing between agency on different levels of analysis and understanding that forms of power are relational and part of particular life assemblages with their own power dynamics.

The case studies in this chapter try to delineate and identify the subjects (collective or individual) of life assemblages that exercise power in the context of genetic testing in China, India and Japan. Here, power in the context of genetic testing is viewed both in terms of discourses of progress and in terms of material and ideological control over the reproduction of the family, including material means of existence, and social institutions and the socialisation of persons. Accordingly, representations of the body form the basis of a number of powerful institutions, including the state, the local community and the family, whereby ideological representations of 'valuable life' of some reign over the views of others, including those of groups of women, disabled and other underprivileged social groups.

Second, in neoliberal versions of Foucauldian thinking in the West, the focus of attention in relation to 'choice' has been shifted away from the state to other forms of abstract power, that is, power regimes associated with modernity and technologies of the self (Kohrman, 2005; Sunder Rajan, 2007; Sleeboom-Faulkner, 2010b). In Foucault's account of biopower, everyone's existence came to be constrained by and dependent upon what biopower's key institutions and discourses ordained as the normal, healthy, competent, well-individuated, non-deviant man, woman or child (Kohrman, 2005: 8). But in such diffuse definitions of power the state has become an invisible constant. In life assemblages, the role of the state cannot be de-emphasised. Rather, its inner workings are crucial to the body politic. People are not just subjected to power relations; they constitute them and give them (knowingly or unknowingly) direction when reproducing them, while authority is shaped in some institutions more than in others. As will become clear in the following case studies, the kind of institutions and socio-political groups in which power resides partly depends on the history of state institutions in a particular country and, partly, on the scope of its subjects for responding to organisational authority at various levels.

Third, Michael Kohrman, anthropologist of China, has argued that corporeality is always fashioned by a fusion of history, knowledge, power and other factors. Yet, the body's role in socio-political formation is always far more than that of a docile discursive object (Kohrman, 2005: 9). Medical anthropologist Margaret Lock (1998) warned against overemphasising mechanisms of power and paying too little attention to the accounts and experiences of individuals in which agency becomes visible. Although confined by social circumstance, resistance is often initiated and orchestrated by individuals, rather than simply being the effects of repressive discipline. Individuals often juggle strategically to their own advantage among networks of power (Lock, 1998: 208). Which factors will be of crucial influence is best estimated in the light of the particular life assemblages individuals are part of at the time. Whether 'biopower' is normalised or naturalised is an open question, as is the logic and inevitability of technological intervention in reproductive dilemmas.

Power is invested both in and outside state institutions, and its modes of production have to be identified case by case. This is so, too, when examining decision-making in the context of genetic testing. To understand individual reproductive decision-making through biopower, there may be a need for more focus on state authority in some countries than in others. Here, reproductive decisions and decisions to acquire 'scientific data' about the genetic make-up of individuals by means of genetic carrier testing cannot be adequately understood in terms of 'power effects'. Decision-making processes regarding genetic testing need to be understood *also* in terms of existing *beliefs* in the creative forces of transcendental authorities, be it a belief in the power of God, the Emperor, biological evolution or nation-state power. Even though Foucault insists that individual behaviour is not entirely determined by 'power', the focus in his work on the mechanisms of power suggests that the practices of daily life are determined by it. The case studies in this chapter more readily acknowledge the complexity and unpredictability of the relationships of couples with genetic testing technology and social and state institutions. Only through intimate knowledge of the regional and situational circumstances of these relationships within the confines of relevant life assemblages can the agency and reproductive decisions of individuals be understood.

In brief, this chapter will examine the effects and consequences of applying a paradigm of Foucauldian biopower onto a context that does not discipline its subjects in a European fashion to plan and shape lives through imported biomedical technologies. The chapter first examines the effects on populations of using genetic technologies in the absence of effective biobureaucratic institutions. And, second, it investigates what effect genetic testing has on communities and individuals that do not discipline themselves sufficiently to become receptive to using new genetic technologies.

Genetic tests can be defined and categorised according to their function, and include carrier testing, genetic screening and prenatal predictive genetic testing. *Carrier testing* is a method used to identify individuals who carry a genetic abnormality that does not affect the health of the person in question, but which increases the risk of producing offspring with a serious genetic disorder. Examples of such conditions are sickle cell disease, thalassaemia and muscular dystrophy (Cases 4 and 5 respectively). Genetic counselling is normally recommended before and after the testing for such conditions in order to prevent confusion over the difference between being an asymptomatic carrier—those who will not develop any signs of the disease—and being someone affected with it. In this regard, Peter Harper (1988) defined genetic counselling as 'the process by which patients or relatives at risk of a disorder that may be hereditary are advised of its consequences, the probability of developing and transmitting it and of the ways in which this may be prevented or ameliorated'. Although genetic counselling aims to minimise adverse psychological reactions, it does not prevent carrier screening from creating opportunities for racial discrimination, as happened with sickle cell screening in the USA in the 1970s (Lee, 2003). Identified carriers can harbour lingering regrets and worries, while others may fail to recall the significance of their test result over time. More constructively, the information provided by carrier tests may be useful in family planning decisions, in the decision to have children, partner choice and in preparing for the possibility of affected offspring (Brandt-Raouf et al., 2006; Prainsack and Siegal, 2006). Considering the potentially serious impact of the outcome of carrier testing and the need for careful consideration of potential carriers and their families, the exploitation of this area of testing by commercial companies has been controversial. For this reason, there is a need to study the practices and motives of commercial carrier screening and the state health sector for offering tests (Sui and Sleeboom-Faulkner, 2007).

Genetic screening is defined by the World Health Organization (1998) as 'tests offered to a population group to identify asymptomatic people at an increased risk from a particular adverse outcome'. Genetic screening, then, is a diagnostic technique applied to a whole population, or to a distinct subgroup within a population, such as newborn infants or pregnant women. Genetic screening can be used to prevent a disease and/or minimise morbidity through early diagnosis and treatment. As screening tests are not definitive, a confirmatory diagnostic test is performed soon after a positive screening test result. This is important to prevent unnecessary anxiety and to enable measures for the prevention or treatment of the condition (Patra and Sleeboom-Faulkner, 2009a).

Although a simple idea in theory, the use of population genetic screening programmes is riddled with bioethical problems that only make sense in the light of the values used and the materiality of the life assemblage they are part of. For example, whether offering such tests is ethical when there are no available methods of treatment or prevention can only be answered by examining the community's life assemblage. Where carried out without appropriate support in place, screening programmes may carry risk implications for the family members of persons screened (Patra and Sleeboom-Faulkner, 2009a). The process of providing information and obtaining consent should be tailored to the level of risk and benefit to the individual and the community with its particular life assemblage. Some forms of pre- and neo-natal screening are provided routinely in countries with high standards of healthcare provision and socio-cultural acceptance of their implementation. In the Asian Pacific area, neo-natal screening programmes for important congenital diseases are well established in some regions, but not so in others. Although the large geographical area and vast ethnic diversity contribute to variation in frequencies of congenital diseases in different regions, some diseases are commonly found in many regions. Examples are the thalassaemias, sickle cell anaemia (SCA), congenital hypothyroidism and glucose-6-phosphate dehydrogenase deficiency.

In general, the bioethical problems associated with the application of prenatal genetic tests (PGTs) are not exceptional compared with other forms of medical diagnosis. Usually, the introduction of predictive/ genetic testing is meant to prevent harm and suffering, and to enable prospective parents and patients to make informed decisions. Yet, concern remains that the introduction of a genetic test is not always based on evidence that the gene(s) examined is (are) associated with the disease in question. Furthermore, it is not always clear that the test itself has clinical validity and is useful to the people being tested (Sui and Sleeboom-Faulkner, 2007). The choice of a genetic test, then, should be based on the individual's best interest and reflect his/her social values (Marteau and Johnston, 1986; Henneman et al., 2001). The use of PGTs also has, usually, unintended effects that may be harmful to individuals and lead to discriminatory practices related to the way life is valued. For example, depending on the life assemblage, the uncovering of genetic information can lead to various forms of exclusion in marriage, school and employment (Case 5), in insurance companies' practices (Porter, 2010), in the inability to access healthcare (Simpson, 2010), in the inability to act upon it, in the moral scruples to abort a fetus with a genetic abnormality (Case 6) or in the unacceptability of giving birth to a handicapped child in the eyes of the community (Scharping, 2003; Nie, 2005).

In European and American moral economies, discussions have been held on the benefits and disadvantages of the availability of genetic knowledge in connection with possible discriminatory practices of employment and insurance, and the provision of medical intervention, follow-up treatment and care when test results are positive. Case studies on genetic testing and screening in Asian local communities show that otherwise seemingly similar moral economies of health have not only radically different treatment implications in wealthy welfare societies and low- and middle-income societies, but also entail varying notions of social and cultural identities of illness, normality and sin. Case 4, on screening and testing among tribal and caste groups in India, provides examples of the consequences of screening for entire populations that are often regarded as inferior and tainted by disease and bad customs. Similar presumptions exist in China about ethnic minorities and rural households. Case 5 shows where a disease in the household is associated with sin and retribution for the behaviour of ancestors. In India, owing to crude screening procedures and porous notions of privacy, examples show that testing methods turn otherwise beneficial screening methods into vehicles of discrimination, with dire consequences for women.

Screening augments already existing social distinctions of gender and socio-cultural status in particular life assemblages. Case 6, which looks at genetic testing in Japan, discusses how socio-economic and cultural histories yield different appreciations of genetic knowledge. In short, this chapter discusses the adoption of screening practices and the availability of treatment in relation to local cultures of blood and inheritance in Asian life assemblages, examining and analysing the material and cultural frameworks that constitute and sustain local moral economies of health. This examination of life assemblages forms the basis for a discussion of the limitations of some approaches to biopower in Asian contexts.

Case 4: Double Discrimination: SCA Prevention Programmes in India[1]
(with Prasanna Kumar Patra)

Sickle cell disease, an inherited blood disorder, is a public health problem for many tribal and rural caste communities in India. Many community members with the condition feel doubly discriminated against—by nature and by the state. This case study examines some of the issues related to the stigmatisation of rural caste communities, who are widely believed to suffer from inherited blood disorders. As shown through this case study, on-going sickle cell control programmes in India confirm stereotypes in their efforts to control the disease.

Sickle cell disease is a blood condition resulting from the inheritance of abnormal genes from both parents. As the sickle cell gene in North and South America, the Caribbean and Europe is usually seen among people of African origin, it is commonly believed to be associated with African ancestry. Similarly, the widely reported high incidence of the sickle cell gene among marginalised groups, such as the tribal communities and 'lower-caste' people in India, causes many to believe that it is an ethnicity-specific disease. In India, many genetic health management programmes initiated by individuals, non-governmental organisations, and public and private healthcare institutions aim to deal with the perpetuation of this sickle cell disease. These programmes include population genetic screening, premarital genetic counselling and medical treatments. This case study discusses the consequences of using sickle cell screening programmes in a situation of socio-political uncertainty and medical service incapacity.

This case study draws on a study aimed at understanding the motivation to participate in genetic research, the level of information received from researchers or data collectors, and expectations from the research. The fieldwork was conducted between December 2006 and May 2007 in three tribal and two caste communities living in four different states in India: Dhodia tribe of Gujarat; Bhil-Pawara tribe of Maharashtra; Sahu-Teli caste of Chhatishgarh and Kondh tribe; and Agaria caste of Orissa. All the communities selected had participated in some kind of genetic intervention programme, such as a carrier screening programme for SCA, and/or human genome diversity initiatives. Interviews were conducted with researchers and doctors working at well-known public and private sector hospitals and research institutes in New Delhi, Hyderabad and Kolkata. No further details on participating communities and individuals can be provided owing to identity protection and confidentiality. Interpreters were used in two community contexts where the participants spoke in their local language/dialect. We do not claim this study to be comprehensively representative of Indian communities or of India in general. However, we do believe that this study represents the general social dynamics experienced by Indian tribal and rural caste communities vis-à-vis bioethical issues in biomedical research. There was a total of 32 interviews (four university teachers/project investigators, six project directors and researchers from various research institutes, 15 community members and seven community leaders), eight case studies and two focus group discussions. Interviewees included university teachers, project investigators, project directors and geneticists. Semi-structured interviews were used to understand the main objectives of the genetic research, procedures followed in subject/community consultation and challenges faced in data collection. Loosely-structured interviews were held with community members and community leaders.

SCA affects an estimated 60–70 million people worldwide and nearly 20 million in India (Angastiniotis et al., 1995; Mandot et al., 2009). Of India's 437 scheduled tribe communities, the sickle cell trait rate (carrier for the disease) ranging from 15% to 20% is found among 20 communities, and it is even higher in certain communities (Basu, 1994). One question that many tribal people ask is why it is mainly tribal people who suffer from this disease. The distribution of the sickle gene coincides with areas of the world where a particularly deadly form of malaria was prevalent. The sickle cell trait has been found to be protection against this particular malaria parasite. Thus, tribal people in India and people of African origin in many parts of the world have a higher prevalence of the disease than that of the mainstream population. Although current studies show an equally high prevalence of the sickle cell gene among several caste communities, the relatively higher frequency and wide reporting of the disease among tribal people crystallises the impression that it is a 'tribal disease', or ethnic-specific. Healthcare provision at a local level often fails to cater adequately to the needs of the community; tribal peoples feel that the reason for such poor provision of services lies in the fact that the upper caste and the mainstream population do not normally suffer from sickle cell disease, and that they are discriminated against at a local and national level. Complaints such as the following are common across affected tribal and 'low-caste' regions:

> You know, as you see it here, it is not a rich or high-caste people's illness. It is not everybody's illness. It is only *adivasis* who get it... (interview Bhil tribal man, 34, Dhadgaon tahsil, Maharashtra; 12 June 2008; transl. from Hindi by Patra).

> We hear that there is not enough money, healthcare and research on sickle cell. It is not a disease of the people who matter. It's a poor man's disease (interview village leader, Sahu-Teli caste community, Raipur district, Chhattisgarh; 24 August 2008, transl. from Hindi by Patra).

In medical science, sickle cell disease is defined as a genetic blood disorder, a disease inherited from both parents. But for many tribal and caste communities in India the aetiology of sickle cell disease and illness behaviour are primarily defined in terms of culture, black magic and superstition. When the genetic nature of the disease is explained to them as part of a genetic disease awareness campaign or a genetic counselling programme, some people find it hard to accept. Rather, they believe the disease to be a malicious attempt by programme officials and jealous neighbouring communities to give their community a bad name. As the reaction of a community member of the Bhil tribe in Dhadgaon district of Maharashtra illustrates:

> We Bhils are the descendants of Rajput [a princely clan/caste], our blood is pure and strong; how can we have any deficiency in our blood? I do not believe in what these people [the medical team] say. They just talk rubbish; they have some ulterior motive, and they want to pollute our community (interview community member of Bhil tribe Dhadgaon tahsil, Maharashtra; 12 June 2008; transl. from Hindi by Patra).

The community leader's reaction illustrates the gap between the views of lay people and experts regarding the aetiology of the disease.

Many sickle cell control programmes distribute colour cards in rural and tribal villages that pictographically show the inheritance pattern of the sickle cell disease gene. Programme managers believe these colour cards are regarded as a useful tool to educate illiterate rural and tribal people. There are two types of colour cards. The 'full yellow' colour card signifies a person as someone affected by sickle cell disease, whereas 'a half yellow/ half white' card indicates that the person is a carrier for the disease. The purpose of such cards is to easily identify an individual, helping doctors treating an emergency to identify the person's sickle cell disease status. But the card is also used for other purposes, such as to regulate marriage: it's a way of avoiding marriage between two carriers through premarital counselling. This is believed to decrease the risk of spreading the disease within the population. While from a public health policy-making perspective the distribution of colour cards makes sense, in a small and closely-knit village where tribal and caste people are bounded by marital and kinship relations, these colour cards have become a means to signify 'deficiency', adding stigma and discrimination to the disease burden.

Genetic counsellors use the colour cards to explain the inheritance pattern of the disease to people as a way of avoiding the same genetic combination in their future offspring. As health officials and programme managers consider premarital counselling as a key preventive strategy, much of the genetic counselling is directive in nature. Unmarried young people and their parents remain the main target groups for counselling. Prospective couples are advised to match their genetic status during marriage negotiations, based on their genetic screening results. As the vast majority of people in tribal and rural areas still follow the tradition of 'arranged marriage', genetic matching based on colour cards plays a significant role. As one genetic counsellor working with the Sickle Project at Raipur in Chhattisgarh said:

> Now we advise people to replace their age-old *janam-kundalis* [astrological horoscopes] with that of *gene-kundalis* [genetic horoscopes]. We believe that this will help reduce the chance of marriage between two carriers and eventually lessen the disease burden on a community. This has been our slogan in rural and urban areas as this is a cultural practice with which people can easily associate (interview genetic counsellor at Dhadgaon Sickle Project site, 24 June 2008; transl. from Hindi by Patra).

In this exercise, unmarried carrier girls are especially subject to discrimination, as the genetic status shown on the colour cards clouds their marriage prospects. Prospective grooms avoid initiating marriage negotiations with a girl with a carrier gene. Although in the absence of adequate medical testing facilities for prenatal testing at local levels such counselling seems acceptable, in many localities this has created a sense of risk among all community members irrespective of their sickle cell disease status or related symptoms. This 'manufactured risk' (Giddens, 1999) has brought the people in these close-knit societies into the ambit of a genetic 'screening culture'.

The impact of colour cards on the marriage prospects of unmarried girls is not the only an issue of stigma and discrimination based on gender. Married women who are carriers are made a scapegoat for having given birth to a diseased or carrier child. Even though their husbands are equally responsible for children with such a genetic make-up, most of the blame is apportioned to the women. Sickle cell sufferers have recurrent crises with severe joint pain that makes them sick and unable to carry out physical labour for certain periods of time. In rural areas where daily labour or physical work is the only source of income, women that became sick become economically dependent on their family. In such cases, they receive little sympathy from their in-laws and family members. Many mothers-in-law complain, and view this as a curse on their family. As one Kondh tribal woman of Phulbani in Orissa said of her sickle cell-diseased daughter-in-law:

> She is an *abhisap* [curse] on my family. She has brought this problem, you know, from her family. She always remains sick. She has not been a healthy wife for my son. She ruined his life. She has given birth to two children who are also sick like her. She has ruined my *vansh* [family line] (interview Kondh tribal woman, of Phulbani district in Orissa; 2 February 2009; transl. from Oriya by Patra).

Such statements are commonly encountered in tribal and rural areas where people have a low literacy rate, a low level of awareness about genetic diseases, have minimal access to basic healthcare and the society is patriarchal in structure.

It is clear that the Bhil were subjected to a sickle cell screening programme by a state that has the biobureaucratic institutions to mediate the frivolous plans of state that does not back up its public health plans enough for them to work at a local level. This has both medical and socio-economic consequences. The disease burden not only affects the social and physical wellbeing of the affected people and the community, but also has a debilitating effect on the economy of the nation.

As the measures taken by the state are sporadic and ad hoc in nature, the 'genetic literacy gap' between lay people and experts continues to cause conflict. Tribes and caste communities experience the cultural set-backs and unmet healthcare needs as destructive.

Those at greater risk for the sickle cell disease gene among tribal and rural caste communities in India believe themselves to be unfortunate on two accounts: first, because there is a 'faulty' gene in their population as a consequence of 'natural selection', and, second, because they are not provided with proper healthcare services by the state to prevent these faulty genes from increasing. Some blame it on God, some on the government and others on both. Many wonder why it is that this disease is more prevalent among people of African ancestry or the tribal communities in India. Without addressing the double disadvantage experienced by sickle cell patients, state policies merely introduce new categories of health, disease and discrimination.

The various beliefs about the significance of being a sickle cell disease carrier are here not fruitfully understood as the truth effect of biopower. Rather, the beliefs in 'natural selection', 'faulty genes', 'karma', 'God' and their signifiers, such as the colour cards and other forms of diagnosis, are imbued with various meanings that reflect current power relations between traditional status differences among castes, and ethnic and gender groups. 'Abnormal bodies' are not so much being managed, institutionalised or classified in terms of biomedical health categories, but denied reproductive life, and discriminated against in terms of already existing gender and caste/ethnic relations. Biopower's currency in the form of 'diagnosis' in this context is not so much productive as oppressive; in this case, it covers up and conceals power relations rather than constituting defining instruments in the production of efficiency, progress and economic expansion.

Case 5: Experiences of Genetic Testing in Chinese Families: Thalassaemia and Muscular Dystrophy[2]
(with Suli Sui)

A presumption of family planning is that the ideal of a healthy and normal family can be pursued. While both practice and science belie the realisation of this ideal, the ideal continues to support a sense of duty to attain it. The result of such family planning strengthens the norm of striving for a 'healthy family' by all. This unattainable norm has augmented fear of genetic discrimination in the family household, making family privacy and confidentiality a matter of great importance. Case 5 shows this to be so with respect to genetic testing for two specific diseases. Prenatal genetic diagnosis is formally regulated by the Chinese Ministry of Health (MoH). According to 'The Measures for the Administration of Prenatal Diagnosis Technology' (MoH, 2003a), only a limited number of certified healthcare providers, available in urban areas, are permitted to offer prenatal genetic diagnosis services. These services are then usually out of the reach of rural communities. Furthermore, the range of tests they provide varies depending on the technology and resources available to each hospital. In Beijing, at present, only five hospitals have permission to offer prenatal genetic testing (Li and Zhou, 2005).

This case study draws on a study of the situation of parents who carry the genes for beta-thalassaemia and their expectations for offspring. Beta-thalassaemia is an autosomal recessive genetic disorder of the blood. In beta-thalassaemia major, haemoglobin production is reduced such that normal growth, development and quality of life can only be achieved by regular red cell transfusions from infancy onward. In this case, thalassaemia means beta-thalassaemia. Data derive from fieldwork examining the circumstances in which parents make reproductive decisions concerning thalassaemia through genetic testing. The fieldwork was conducted in Nanning, Guangxi Province, and Chengdu, the provincial capital of Sichuan. As part of fieldwork in hospitals and research institutes, Suli Sui interviewed eight families with thalassaemia-affected children, four geneticists, two genetic researchers, two haematologists and insurance company agents. Appropriate permission was obtained from the hospitals and research institutes.

Thalassaemia

In contrast to other parts of China, thalassaemia (including alpha- and beta-thalassaemia) is widely prevalent in the south of China, especially in the provinces of Guangxi, Guangdong and Hainan. According to the Family Planning Committee of Nanning, in Guangxi Province 20% of the population carry the thalassaemia gene, while Guangdong Province has a prevalence of approximately 12% and Hainan a prevalence of 15% (Li, 2006). Thalassaemia is a recessive condition, and carriers are usually not aware of having the gene without taking a test, which is available mainly in large cities. When premarital check-ups were still compulsory, some cities, such as Nanning and Guangzhou, screened automatically for thalassaemia. But, after 2003, when check-ups were made voluntary, the number of couples taking premarital tests dropped sharply. To stimulate participation in the testing programmes, the governments of Guangzhou and Nanning began to offer free tests, also targeting couples in rural areas (Li, 2006). According to the 'Guidelines for Genetic Counselling' (MOH, 2003b), genetic counselling should be offered by clinicians who have a background in genetics. But, in practice, they are usually paediatricians and obstetricians. Many of those seeking a carrier test and counselling have previously had an affected child. The genetic counsellor (i.e. the paediatrician or obstetrician) advises the family to take a genetic test to confirm the carrier condition of parents and child. According to the family planning policy, parents of a seriously handicapped child are allowed to apply for permission to have a second child (Sui and Sleeboom-Faulkner, 2010a). The availability of genetic testing, then, is crucial for couples in their decision to give birth. Without the tests, the only acceptable alternative to the carriers is not to give birth and opt to adopt a child. In many cases, however, couples use abortion to terminate thalassaemia-affected fetuses. This is in line with the family-planning policy, which advises doctors to suggest that the pregnant mother should have an abortion in cases of severe disability (interview Life Scientist F).

The usual medical treatment for thalassaemia patients is blood transfusion. To a great extent, the life of an affected child depends on the blood transfusion, but for many families the costs of blood transfusion are very high, for instance approximately two payments of 2000 renminbi (RMB) (c. £150 in 2007) in one month at the haematology department of 303 Hospital (Sui and Sleeboom-Faulkner, 2010a). Thalassaemia in China is a great economic burden for the majority of families with the disease, especially for poor rural families, many of whom are forced to discontinue treatment. Blood transfusion cannot cure the disorder, and the prospects and life expectancy of thalassaemia-affected children is unclear. To cure an affected child, some couples decide to use a subsequent unaffected child's umbilical cord blood—if a tissue match—to save the thalassaemia-affected sibling. Although the idea of creating a 'saviour sibling' for the purpose of saving a sibling did not receive much attention, the high cost of umbilical cord blood transplantation, approximately RMB 200,000, did.

If the result of the prenatal genetic test for thalassaemia is positive, terminating the pregnancy is the usual, and socially accepted and preferred, choice. If the test is negative, but the tissue of the fetus does not match that of the thalassaemia-affected sibling, the pregnancy will also be terminated, as its umbilical cord blood would have no use. Owing to the one-child policy, giving birth to this child would mean no more chances to create a 'saviour sibling' with a match. In practice, prenatal diagnosis requires amniocentesis, a procedure associated with risk for the pregnancy, which often takes place around mid-term. Abortion in such cases is an especially painful decision for the mother. One woman interviewed decided to conceive again in order to save the life of her son:

> As a mother I try to give a chance to my child to live. A nearly five-month-old fetus already has a 'person shape' [*you renyang*]...To my son, I am a kind mother, but I feel like I am also a murderer at the same time (interview Yin, 19 February 2008; transl. by Sui).

It had been hoped that umbilical cord blood banks that have been set up in China recently would offer better prospects. According to statistics released by the MoH, the number of blood samples in stock in 2007 was approximately 25,000; among them, 400 samples have been used in clinical transplantation for leukaemia patients (Wei, 2008). However, there are still no cases of matching samples in the umbilical cord blood banks for the use in transplantation for thalassaemia patients. This situation limits the possibilities of the parents of thalassaemia children, continuing the complex reasons for choosing 'saviour siblings'.

It is clear that the state's population policy set the narrow parameters for the number of children a couple may have, thereby creating a situation in which healthy fetuses are produced solely to save the life of a thalassaemia-affected sibling, while thalassaemia-affected fetuses are aborted. Available means conducive to the amelioration, cure or prevention of the disease, such as blood transfusion, bone marrow transplantation or pre-implantation genetic diagnosis are out of the financial reach of the poor. Although impoverished parents seem to make rational decisions in their situation, their emotional health is damaged in the process. For this reason, questions can be asked about the 'productive' nature of the population policy. If multiple pregnancies are needed to produce a 'sufficiently' healthy child, the one-child policy, in combination with technological possibilities in China, has caused the normalisation of health to take on perverse forms.

Duchenne Muscular Dystrophy

In China, about 650,000 boys have Duchenne muscular dystrophy (DMD) (CRCF, 2005). DMD, a common childhood muscular dystrophy, is a lethal X-linked genetic disorder (passed on by the mother) usually involving a gradual deterioration of muscle function, and affected individuals require long-term care (Gardner-Medwin and Sharples, 1989). Testing for DMD is aimed at early diagnosis, primarily to offer genetic counselling and, where possible, prenatal diagnosis. When diagnosis of DMD takes place, the family is warned that DMD is an inherited disorder and that the sisters and cousins of the affected boy have a high risk of being a carrier. Improving the identification of carriers is aimed at reducing the number of boys with DMD being born into affected families, as is largely the case in Europe (Emery, 1991).

In traditional Chinese reproductive views, women are generally regarded as responsible for a healthy birth, and boy preference is salient. Only the male is treated as family offspring, which is called 'burning incense' (*xianghuo*) (Liu, 2005), for traditional custom prescribes that only a son can hold a memorial ceremony for his ancestors. Knowledge about the X-linked inheritance of DMD may result in the blaming of the mother, as it is passed on genetically in the female line and usually occurs in boys. In a counselling clinic for DMD, doctors interviewed say they try to avoid the issue of who is to blame (interview 17 October 2007). During genetic consultation, the doctors explain some basic knowledge of DMD, without mentioning the X-linked inherited character directly (Sui and Sleeboom-Faulkner, 2010b).

Even when informed that there is no effective medical treatment for DMD, many parents try to find an alternative treatment. Some parents beg the doctor to use their son as a human subject in clinical experiments on DMD. Other families turn to traditional Chinese medicine, even though its benefits are doubtful (Urtizberea et al., 2003). Parents tempted to access drugs are often not informed of the effects and risks associated with the taking of the drugs, although the potential toxicity of some of these traditional Chinese medicines has been clearly demonstrated elsewhere in other conditions (Chan and Critchley 1996; Critchley et al., 2000). Some hospitals, in exaggerated advertisements, boast about special treatments for incurable diseases such as DMD (Chen, 2005). The advertisements attract desperate people, including those from remote rural areas. The strong desire for treatment easily misleads DMD-affected families into spending money, worsening their financial conditions (Sui and Sleeboom-Faulkner, 2007).

The long-term and deteriorating nature of DMD creates an increasing burden on families. Family members must provide increasing assistance with activities of daily living, while long-term care needs usually increase the financial burden on the family, particularly if one of the parents has to forego employment to care for the child (Chen, 2005). A popular view in China considers severe illness as punishment for an ancestor's misbehaviour or for the family's current misconduct, calling it 'religious punishment' (*zongjiao chengfa*) or 'pre-existence retribution' (*qianshi baoying*) (Wang and Zhang, 2002). This popular explanation stigmatises families with severe handicaps. Leading a 'normal' life is difficult for those regarded as disabled, as it is hard for them to find a partner, and as their disability isolates them from the community (Kohrman, 1999). One woman, Ms Lu, expressed this sentiment:

> The symptoms of my son are becoming obvious and we cannot keep it secret, but we do not say more about this illness to others except to our very close relatives. We are afraid of gossip. We are very cautious and scrupulous to avoid offending anyone. In school, several classmates laughed at my son and he feels bad. The only thing we can do is to implore the teacher to pay more attention (interview Lu; 27 November 2007; transl by Sui).

Similarly, families with a son lost to DMD feel stigmatised. One mother, Mrs Yang, expressed fear for her family:

> A family without a son is despised in this village. Some people think our family is cursed by a devil. In our village, usually, a matchmaker will come to a girl's home to propose a candidate for marriage [*tiqin*] when she approaches the age of twenty. My daughter is twenty-three, and no matchmaker came for her (interview Yang; 7 December 2007; transl. by Sui).

Matchmakers, then, are not interested in women from DMD families, as they are believed to produce inferior offspring. DMD family members interviewed in this respect expressed similar views on their isolated position and prospects in society. In China, there are several Internet communities for 'muscle illness patients' (*jiji huanzhe*). Communities, such as 'going with you' (*jingcai tongxing*) and 'window for muscle atrophy' (*ji weisuo zhi chuang*) offer virtual space for patients and their family to communicate and share their experience (cf. http://www.jingcai. org and http://bbs.39.net/by/forum/1324-1-1.html) (Sui and Sleeboom-Faulkner, 2010b). Although helpful to DMD patients and their families to some extent, most rural areas have no access to, or knowledge of, the Internet.

China homes predominantly patrilineal communities, where son preference is a main reproductive strategy. In this context of son preference, female DMD carriers have little or no reproductive value, and are rejected for various reasons. First, they are blamed for passing on X chromosome-related diseases; second, only their daughters are not likely to be symptomatic for DMD; and, third, their daughters may also be DMD carriers.

This makes female carriers reproductively futile, affecting the standing of the entire family household. Not surprisingly, families go to extremes to find cures, which may include unproven therapies and fame medicine. From the point of view of biopower, DMD carrier testing has ambiguous meaning: it can decrease the number of DMD cases, which may mean a more healthy and productive population, but it can also devalue the reproductive abilities of women of entire households, wreaking havoc on lineage, and hindering the healthy reproductive lives of families with DMD.

Case 6: Genetic Testing and the Family in Japan[3]
(with Masae Kato)

Although certified clinical geneticists (CCGs) are careful not to generalise about patients' motivation for taking a genetic test, interviews with patients indicate that in Japan the issues of marriage and reproduction are outstanding concerns and the main reason for taking a genetic test. These two concerns are visible in the stated motivations of young adults taking a genetic test. A number of studies (e.g. Numabe, 2007), as well as official documents (MEXT, 2000), have emphasised the importance of respecting culture in dealing with a genetic disorder and genetic information. Respect for culture is, of course, important. Yet, what seems to be respect for culture might sometimes actually be hiding superstitions and culturally embedded prejudice about genetic disorders. This case study shows how genetic tests can augment cultural taboos through the ways in which 'genetic information' is understood in the socio-cultural context of Japan.

This case study draws on a field study conducted in Japan in 2006 and 2007, in which 12 patients from 12 households were interviewed about their experiences with genetic testing. Patients were contacted on a personal basis. They could not be approached through official institutions because of issues of confidentiality and privacy. Six out of eight households took a genetic test; two did not. All patients gave their free consent to take part in this study, but their names have been replaced with pseudonyms. The interviews showed that the decision to take a genetic test depends largely on the interviewee's views of marriage and plans for reproduction. In support of the analysis of the eight cases, this article refers to the evidence from 46 other interviews.

At-risk people and people with symptoms decide to take a genetic test during three critical occasions in the course of life: (i) when choosing the direction of their studies (*shinro*); (ii) when deciding on a job (*shûshoku*);

and (iii) before marriage (*kekkon*). Choosing which path of study to follow and job hunting are important because the symptoms of muscular dystrophy are not compatible with jobs requiring nimble fingers (Tomiwa, 1998: 55). While children are still of a young age, parents of at-risk children only tell them which jobs are *not* suitable for them. Mr Hara, who has DMD, is now studying computer systems because his mother, Ms Hara, suggested this area in the knowledge that he would not be able to manage a full working life. However, the time of choosing his direction of study did not prompt him to take a genetic test. Only when marriage became a realistic possibility did he feel sufficiently motivated to take the genetic test.

Marriage in Japan was, and still is, a union of two family lines rather than between two individuals (Yoshizumi, 1995: 188). Under the Civil Code, women usually enter into the family line of their husbands. Among interviewees, marriage to a person with a genetic disorder was regarded as a scandal for the family of the spouse. A person with a genetic disorder is seen to contaminate the blood of the other family line. Ms Kamisawa, who has a genetic bone disorder, expressed that she was called a devil (*akuma*) by the family members of her ex-boyfriend and said that her existence in itself is a curse (*tatari*). In her current partnership, her father-in-law would not accept their marriage initially. But, with the birth of a grand-daughter, he finally accepted their marriage, but only after he converted to Christianity. Ms Kamisawa assumes that this enabled him to accept a sick woman as a daughter-in-law, and to expiate his sin, which resulted in his son marrying a woman with a genetic disorder.

A person with a genetic disorder entering another family is considered as 'contaminating blood' and as 'giving shame' to the other family line. This mindset was found more among grandparents, but also, to some extent, among younger people, albeit less, and especially among couples with a genetic disorder. Among 46 cases, family members regarded people with cerebral palsy less harshly. Some family members said: 'I accepted their marriage because the disorder is not genetic' (*iden ja nai kara*). But why feel offended if a child with a genetic anomaly is born into your family when the child is not your own? Anthropologist Margaret Lock's account of health and family is relevant here:

> In Japanese society the inability to perform one's role in a given group adequately generates a sense of not only having let oneself down, but of having let down the group as a whole. This can stimulate strong feelings of guilt, whether the group is one's family, co-workers, or others (1980: 248).

In this case, the role of a couple, especially the woman, in a given group is to reproduce a healthy child for two families. The birth of a child with an anomaly is thus not only a failure of the young parents to live up to the expectation of the two family lines, but also a failure to safeguard the honour of the two families. Among the other 46 cases, it was also found that family members started to undertake religious practices upon their family member's marriage to a person with a genetic disorder. When a genetic anomaly in a newborn appears spontaneously, the question of 'which family line is guilty' becomes a concern for family members. Dr Nakamura (male paediatrician, Tokyo) said:

> It is interesting to see how members of the two families sit during the counselling sessions. Members of the family of the affected spouse sit looking down on the desk, while members of the other spouse sit straight...In my experience, it is rare that spouses lose trust in each other as a result of the discovery of disorders. It is more because of continuous grumbling of parents-in-law about the disorder that couples lose harmony and divorce in the end (interview Nakamura; 22 June 2007; transl. by Kato).

Although there are no official statistics to show the percentage of divorces attributable to the discovery of a genetic disorder, CCGs say that the divorce rate is twice to three times higher among couples with genetic disorders than among couples without. Ms C, a carrier of DMD, was divorced by her husband. Before he left, he accused her as follows: 'Why didn't you tell me that you are a carrier before we got married? You brought contaminated blood [*warui chi*] into our family'. In Japanese society, marriage is almost synonymous with procreation. It is reported that in Japan, in 2005, 96.4% of married couples had at least one child (MoHWL, 2005). Procreation is a means of perpetuating the family line. Marriage, then, becomes problematic where there is a chance of transmitting a genetic anomaly to the next generation.

Those who have given birth to children with a disability often say they feel guilt about having failed to give the best to their children and about being the cause of their child's pain. A Japanese scholar in the study of nursing, Tsukamoto, cites a narrative from her interview with a mother whose child had a spontaneous genetic anomaly:

> The issue of the child's anomaly is beyond our control. It just happened. Yet, my child came out of my body, and I, as a woman, must take full responsibility for my child for the rest of my life...My child

came out of my body. I cannot blame myself too much. All is my fault. My child took bad parts of mine and came into existence. I thought that I was healthy, but I could not give my child a healthy life...Of course I blame myself (2005: 182–3).

Genetic counsellor Tamura[4] notes that the fact that their children will fall genetically ill in the future is a great shock for many of the clients (2007: 77). When a couple does not have a child yet, they are told by people close to them 'Would you knowingly make your child suffer?' (*shittete kurô saseru no ka?*) (Cases 1, 2, 3, 6 and 8). In addition to this, a genetic disorder is thought to be absolutely incurable, for which they are intensely disliked (*imikirawareru*). Mr Honda was diagnosed with spinal muscular atrophy in 2001, which is a late-onset muscular disorder in men. Being asked what his first reaction was after the diagnosis, Mr Honda explains:

It was shocking to know that my disorder was genetic. It means that this sickness is myself (*jibun jishin ga byôki nanda*) and it cannot be cured at all...I cried for three full days, wrapping myself in a blanket. Having this information, I felt that I was branded (interview Honda; 23 May 2007; transl. by Kato).

Mr D continues to say that although his physical condition is increasingly weakening, he is not worried about his life any more, as he is over 60 years of age. He worked until he retired, did everything he wanted to do and his children are independent. His only worry is the future of his male grandchildren, as he has only daughters. He said he is sorry about the fact that they might have inherited the same disorder.

All the CCGs report that although clients might raise technical questions about the treatment in the first part of counselling sessions, in the end most of the clients express self-blame (*jiseki no nen*), sometimes bursting into tears at the thought of having transmitted their disorders to their children. CCGs also report that some of their clients told them that when they have quarrels with their children, the children blame their parents for having transmitted a genetic disorder, and even for having given birth to them.

How do cultural notions of 'care', 'family', 'disorder' and 'independence' contribute to a culture of intensely disliking hereditary genetic disease (*idenbyô*)? More than half of the individuals out of the 46 cases studied expressed anxiety about receiving 'care' (*sewa, mendô*). For example, worries about becoming bedridden, taking care of oneself and being dependent were expressed by most of the 46 individuals. CCGs also mentioned that the first question from clients during the counselling sessions is whether they will need extra equipment to survive, and if they will need extra care from others. From these accounts, it is clear that dependence is intensely disliked. If someone is certainly going to lose independence, the question of who will take care of him or her arises.

The culture of family and care in relation to the intense dislike of genetic heredity (*iden*) is role-specific in Japan. Fear of the child becoming a burden to the family was expressed in five out of eight cases. Ms Ikegami said:

> Who will take care of my future children when they start dialysis? There is no treatment for this disorder. So, I do not give birth to a child. To take care of myself is enough (*jibun no koto de sei ippai*) (interview Ikegami; 17 September, 2007; transl. by Kato).

Some parents seemed to be preparing for the future of their children, after they lose independence. Ms H said that her younger sister, who doesn't have a disorder, was raised with the idea that she had to take care of her older sister, Ms H, for the rest of her life. She was also told not to get married. Among the other 45 interviewees, similar statements were found. For example, one son was told to find a partner good at cooking so that she could take care of his sister, who had a disability (there were four similar cases). In five cases, a sister of a disabled sibling was sent to a school to learn household chores to prepare for the future. Females are more often expected to provide care. Parents of disabled children stated that they gave care out of love and affection (*aijô*), and with the idea of compensating for the illness they had given to their children (*oya toshite no sekinin de oginau*). Parents' guilt about transmitting a disorder is explained by various factors, and one of the prominent explanations is the fact that they transmitted a life of dependence to their children.

The empirical studies of Traphagan and Kiefer on the elderly, senility (*boke*) and family care in Japan provide helpful analyses because such problems and problems of genetic disorders have much in common: patients are 'less able', 'health problems tend to be multiple' and 'their capacity to recover is impaired' (Kiefer, 1987: 104). Traphagan points out that elderly people try to avoid seeing their forgetfulness as 'biomedical forms of dementia' (1998: 94), such as Alzheimer's disease, which cannot be cured. Instead, they try to see it as something caused by exterior conditions, such as age, environment and a lack of daily exercise—as something that can be cured in the end (ibid). Incurable biomedical disorders are feared, as they might require continued care from family members (ibid).

Although the situation in Japan appears to be changing, in many Japanese families, particularly in rural areas, only one child receives the majority of inheritance and takes on the position of head of the household. This comes with the expectation that the inheritance will be reciprocated in the form of provision of healthcare, a place to live and economic support as needed in old age (Traphagan, 1998: 93; Long, 2005: 90). Older Japanese people are often able to legitimately expect to depend on at least one child for social support in old age. There is a delicate interdependent relationship among family members regarding the issue of care, and to become senile means placing an undue burden on family members.

Senility is feared because it involves 'undue' and illegitimate demands in terms of family interdependence, which disturbs the balance of interaction within a household. It means that the aged fall into a position where their behaviour is at odds with the needs of the group—specifically, one's family (Traphagan, 1998: 93–5). To avoid this, efforts are made by the aged not to become senile, so they can legitimately remain in the system of family care. Genetic disorders share similar aspects. There is a deep-rooted idea that a genetic disorder is incurable, as Mr D's account indicates, and, on top of the physical pain and suffering, there is a fear of losing one's independence, as well as becoming a burden to the community. Fear of genetic disorders can be understood in terms of fear of becoming an undue burden and a fear that their offspring will become an undue burden to their family and society. Genetic information becomes particularly sensitive and is not easily shared, as parents do not want their children to worry that they may become a burden to the family in the future (*shinpai sasetaku nai*), or that their children may themselves give birth to a child with a genetic disorder (Nomaguchi et al., 2007: 57).

One might wonder if the public healthcare system can solve the problem of this burden. Since April 2006, the 'law to support the independence of disabled people' (*Shôgaisha jiritsu shien hô*) was introduced (MoHWL, 2006). According to this law, protection from the state in the form of financial support was to be reduced, including assistance for people with a genetic disorder. For example, a person who uses public health services now pays 10% of its cost, while before 2006 more than 90% of the costs were covered by the state. This policy is now applied to all disabled people, regardless of their income, while before 2006 those with low social security benefits did not need to pay for the service at all.[5] The name of the law implies that it supports 'the independence of the disabled', but, in fact, it withdraws the state's care that enables people with a disability to live independently. Less support from the state means that more support is needed from family members. All the interviewees in this study expressed their anxiety about their standard of living in the future, and worried about becoming an undue burden on their family.

In brief, because in Japan the idea of a successful marriage is strongly linked to notions of reproduction, genetic testing may, in effect, endanger a person's chances of sustaining a marriage and having offspring. Genetic disorders are perceived to contaminate the blood of the family lineage and arouse a sense of shame, guilt and fear in affected families. Family roles in Japan are associated with types of care provided and received, which includes specific expectations of what it means to be independent. Fearing the increasingly unsustainable demands placed on the family household, affected members keep the test results secret from other family members. So as not to worry ageing or affected family members about the disease, doctors and family members keep the information from the patient as long as possible.[6]

In the past, the Japanese state has provided generously for genetically afflicted individuals opting to live on their own. But the fate of affected individuals increasingly lies in the hands of the family, as state policies have decreased support for living arrangements encouraging the independence of disabled people. Rather than supporting the autonomy and decision-making power of disabled individuals, the 2006 'law to support independence of disabled people' resembles UK prime minister David Cameron's 'Big Society' idea: it stimulates the independence of individuals by rendering them dependent on the local community through financial cuts. This is a great burden for this low-fertility and ageing society where, increasingly, both parents need jobs to sustain the family, including both sets of grandparents.

Conclusion

This chapter examined the effects and consequences of applying genetic technologies among populations and their diverging life assemblages. In terms of biopower, biocitizens are expected to make rational decisions to maximise their life value. But, in the absence of effective biobureaucratic institutions, other forces than biopower are exercised in the Asian life assemblages studied. In the light of particular life assemblages in Indian tribal communities, in China and in Japan the logic and inevitability of technological intervention in reproductive dilemmas was questioned, as was the notion that the application of new technologies automatically has a productive effect. Following the logic of SCA screening programmes among tribal communities in India, it was not the rationality of using the screening programme in the tribal societies, but the meanings that were attributed to carriership and its consequences for marriage prospects and family reputation that was important to individuals, families and communities. The case study of genetic testing for thalassaemia and DMD in China showed that, under its one-child population policy, the creation of 'saviour siblings' was normalised among carrier parents who had no other options open to them to save their affected child. Although this choice might seem 'rational' and 'productive', it is accompanied by extreme emotional and physical suffering, and often failure. Although DMD testing may help identify and prevent cases of DMD, in China it also can endanger reputation, and therefore the reproductive lives and livelihood of entire family households, especially of women. If the cultivation of a healthy population and workforce would signify the productiveness of biopower, then DMD testing is failing on the account of both public health, as it leaves behind stigmatised families with few reproductive options and as it instils fear into potential carriers, and individuals, who suffer as a result of their 'rational choices'.

Whether the rationale of genetic testing clashes with or reinforces reigning notions of identity in traditional cultures cannot be adequately comprehended through the one-sided application of notions of managing population health and the productivity of biopower. Rational biomedical arguments may play a role in both the decision to undergo genetic testing and in reproductive decision-making. But, in practice, people make decisions while fighting fear, and amidst cultural givens that prevent the person from acquiring information and making autonomous decisions. In the case study of SCA in India, we see that genetic screening is deployed as a tool in intertribal battle; in the case study of DMD in China and genetic disease in Japan, we see taboos on genetic diseases set off by the knowledge of cross-generational effects; and in Japan we also see the hiding of test results and genetic information from family, community and from the carriers themselves.

Although genetic screening tests may have been intended to enable communities and individuals to base reproductive decisions more rationally and efficiently on genetic information, we also see that 'scientific' results were adapted to and interpreted through a particular life assemblage in terms of colour cards, bad and good blood, and household status, which shape local conditions of discrimination, stigma, the need for 'saviour siblings', and the hiding of genetic information. Any strategy that divorces practices of genetic screening and genetic testing from the values attached to it and the capacity to deal with it as intended, makes use of the ideology of scientistic universalism: it is a belief that scientific knowledge tools can enhance any society without the evidence to support this. Knowledge of life assemblages could prevent such misinterpretations by thinking through the consequences of community life interventions beforehand.

4
Human Genetic Biobanking and Life Assemblages in Asia: Transnational Moral Economies of Health, Progress and Exploitation

The nature of the relation between medical research and society is changing in fundamental ways. The focus of medical research has moved away from the study of individual bodies as physical entities to medical studies based on networks of data, information and archival systems, such as biobanks. A shift is taking place, then, from the hospital as the site of biomedical research to the laboratory. At the same time, the management of public and individual healthcare has come to depend on technologically advanced systems for maintaining healthcare data, and blood and tissues in biobanks, requiring new methods of data collection, maintenance and management. A seemingly contradictory process is occurring in the management of healthcare data and tissues. On the one hand, body tissues and data are processed, stored and standardised so as to become applicable to broad categories of potential patient populations. On the other hand, data and tissues are re-assembled to fit the specific individual in tailor-made medicine (Waldby and Mitchell, 2006). Biobanks, then, are central to the organisation, maintenance and manipulation of data and tissues from laboratory and hospitals. This chapter focuses, in particular, on the collection of medical data and tissues in biobanks, and its consequences for different life assemblages.

A biobank can be used for the collection, storage, maintenance and application of genetic samples and information for various purposes. The information in human genetic biobanks affects the organisation of social life and medical possibilities in many ways. For instance, it facilitates the expansion of research efforts in the fields of forensic science and epidemiology. Thus, the police use forensic genetic databanks to match DNA found at the scene of a crime, and genetic biobanks can be used in studies aimed at determining the relationship between disease and the genetic make-up of individuals and groups. This relation, again, can be matched with ethnic background, living environment and socio-cultural group membership. The creation of genetic biobanks also makes possible the collection and management of personal information, the disclosure and control of which has far-reaching social, political and financial consequences. The disclosure of the genetic make-up of an individual with a serious genetic defect, for instance, may have considerable consequences for an individual's family relations, employment, insurance possibilities and life expectation. When a genetic defect is linked to a group of people or lifestyle, the consequences may be even more far-reaching and lead to socio-genetic discrimination. Human genetic biobanking, then, is of great concern to society, and has been the subject of discussion in the social sciences and various countries, but mainly concentrated in European countries and in North America (Knoppers, 2003; Sándor, 2003; Sustrop 2004). These discussions, however, are held as if the questions to be dealt with are basically the same and vary only to the extent that different countries may have divergent political preferences.

Political preferences in Europe are usually explored through public consultation and are partly meant to educate the public about the use of biobanks in science and in social life. This is of particular importance to regulatory authorities, as the success of biobanks depends largely on the cooperation of the public. The creation and management of biobanks, then, calls for sensitive policies regarding the regulation of biobanking activities. Biobanking and regulation are shaped differently through their life assemblages. Variation is largely owing to the existence of diverging systems of socio-political and economic organisation, different functions of and attitudes towards science, and different ways in which governments relate to their populations. As policy decisions about biobanking in many countries have shifted towards the nation-state and nation-state governance, as well as other local units of decision-making, this constitutes their life assemblages. But, as will become clear, life assemblages are influenced and co-shaped in relation to and in negotiation with a wider world in which biobanking and the 'harmonisation' of regulation evolve.

The collection and storing activities of human DNA banks serve various areas of research. As part of the Human Genome Project (HGP), which was initiated in 1991, these forms of data collection served the 'completion' of the human genome map in 2001 (Venter et al., 2001; Lander et al., 2008). While some scientists study the 'average' or 'standard' genome, others study variations within and between populations. An increasing number of projects now focus more directly on human variation and diversity by sequencing the DNA of selected, and supposedly more genetically 'pure', 'indigenous populations'.[1] Detecting genetic variation among such populations becomes easier if there is a generic sequence for comparison. And as the HGP has 'completed' the map of the 'average' human genome, there has been increased interest in sampling indigenous populations in order to study human diversity. Population geneticists hope that identifying differences in genetic sequences between people will help determine what makes groups of people different from other groups. Of course, other factors also contribute to population differences, such as culture, language, environment and lived experience. Although in some population studies these factors have been hard to estimate (e.g. Majumder, 2000), it is expected that the identification of genetic difference between ethnic groups (such as a different gene sequence that causes immunity to a disease) will have human medical, cultural and historical, and economic value. However, this perceived value in the 1990s also resulted in a 'gold rush' where universities, governments, corporations and private researchers were seeking to identify human genetic variation (Marshall, 1997: 565).

Another international effort, the Human Genome Diversity Project (HGDP), started earlier than the HGP and involved genetic sampling among indigenous populations. According to Luca Cavalli-Sforza, its initiator, the HGDP served to avoid the irreversible loss of precious genetic information (Cavalli-Sforza et al., 1991: 490–1; Chu, 1998). Referring to indigenous populations as 'isolates of historic interest', the HGDP planned to immortalise the DNA of disappearing populations for future study (cf. HUGO (Human Genome Organization) Committee, 1995). In this context, one aim of the creation of gene banks for scientists was to reconstruct the history of the world's populations by studying genetic variation to determine patterns of human migration.

However, the initial conceptualisation of the HGDP has been widely criticised for its consideration of indigenous peoples as mere research subjects, with little regard for their continued livelihood. Various ethnic groups have protested against population genetics, which they claim has been conducted without prior consultation and without consulting the individuals and communities involved (cf. Peterson, 2001; Pilnick, 2002; Knoppers, 2003; Sándor, 2003; Sustrop, 2004). Because the HGDP was planning to collect blood samples, some groups called the HGDP the 'vampire project' (Lock, 1994: 603–6), while others were angry because they believed that they were possible target populations, even though no community representatives had been contacted about the project.[2]

In some low- and middle-income countries with great ethnic diversity and little access to healthcare, such as India, China and Indonesia, attempts have been made to sample the populations and map genetic resources cheaply. Such populations are targeted because they are thought to be genetically isolated and, in some cases, have well-documented genealogical histories, making them valuable for tracing inherited diseases. In Asian tribal and isolated communities, which are the likely subjects for diversity studies, living in remote areas means people are often illiterate and suffer from nutritional and parasitic disorders. Here, it may be important to distinguish between studies that are aimed at improving the healthcare of the local community and those that aim at sampling isolated communities to study the diseases of the wealthy, such as diabetes and coronary heart disease. However, in practice, the collecting of a great number of genetic populations is hoped to increase the chances of finding useful information about the DNA responsible for syndromes.

In analogy with the enclosure of agricultural land, entitlement and property of the genome is now spoken of as the 'new enclosure'. Especially in transnational efforts to make biobanking a lucrative activity, such enclosures entail complex bioethical issues related to genetic sampling and the use of biobanks in different localities. Many discussions on biobanking in Western countries presume that setting up population banks and cohort studies will ultimately serve the health of the population. For this reason, an appeal is made to the goodwill of national and local populations to contribute bodily tissue. At the same time, explanations are provided to the public as to what research targets are relevant to the biobank's research, and agreements are written up about the feedback of medical information on the sampled person and his/her family if relevant disease information is found. Explicitly expressing concern for the protection and interests of donors, procedures for informed consent taking for biobanks seem to symbolise one of the global moral economy's pillars in the life sciences.

In this chapter the three case studies on human genetic biobanking in Asia aim to illuminate the life assemblages conditioning genetic sampling in domestic biobanking, transnational science collaborations and socialist moral economies. The first, Case 7, on the Indian National Repository, concerns the considerations taken into account in the planning of India's national biobank. For the National Repository to be valuable as a resource for scientists who need to publish internationally, the data need to conform to international sampling conventions, which include the taking of informed consent. Fieldwork in India provides an indication of the difficulties this poses for existing and future sampling practices among tribal and small-scale communities. Case 8, on the human genetic biobanking efforts of Dutch and Indonesian scientists, looks at the two forms of transnational collaboration that underlie biobanking efforts, examining the different subjectivities associated with the national background of the scientists concerned. This case study sheds doubt on the application of simplistic paradigms of postcolonial exploitation and of scientific progress. Case 9, a study of a large-scale biobanking effort in mainland China, describes the change of biobanking plans under the influence of the international community, indicating that the image of a biobanking project can greatly alter the international and local reception of genetic sampling activities. The case study illustrates the need to look beyond formal guidelines that serve the requirements of a global science community at local sampling practices.

Case 7: Setting up the National Repository in India: Ethicalising Human Genetic Samples[3]
(with Prasanna Kumar Patra)

India has an abundance of biomaterial collections, although exact numbers are not available. In a 'landmark' initiative spearheaded by the Anthropological Survey of India, the Indian government set up a national advisory committee for the 'establishment of a permanent National Repository for human genetic resources and data' in 2006. The National Repository is considered to be a landmark achievement in terms of managing the human genetic diversity of the country (AnSI, 2006). This initiative started with the idea of providing a centralised repository to the randomly and unsystematically collected and stored biospecimens and health data at various research centres, including those of universities, public and private hospitals, and research institutes. The endeavour of establishing a permanent national repository for human genetic resources and data in India is a by-product of developments in other fields. The last one and a half decades have seen India as an emerging economic power. After liberalisation of its economy in the early 1990s, it has maintained a high and steady growth rate of about 7–8% of its gross domestic product (GDP). It acquired most of its leverages from service sectors such as information technology (IT), which has spawned a burgeoning bioinformatics industry (Acharya et al., 2004). Indian biotechnology companies play an important part in the innovation of new technologies on a global level. In 2003–04, Indian biotechnology companies had combined revenues of more than US$700 million and they surpassed US$1 billion in 2005. According to the World Health Organization (WHO), India is the fourth largest producer of pharmaceuticals (Jayaram, 2005). This has generated investment in research and development in every field of science, including the life sciences and genomic research. 'Public–private partnership' is a new buzz phrase in this field, and several public genomic research institutes of repute have set up collaborative research programmes with leading national and multinational pharmaceutical and IT companies in throughput research areas (Anonymous, 2002).

An increase in funds and improvement in infrastructure has pushed many research institutes to take up genomic research with vigour. In 2001, the Department of Biotechnology negotiated with pharmaceutical and biotechnology companies for collaborative ventures in secondary genomic research, and the focus of these projects was in areas such as proteomics, functional genomics, data mining, and curation and annotation of genome data. The biotechnology boom, coupled with India's rich biodiversity, a huge medical infrastructure, a large population size and an expanding market, has made India a hunting ground for genomic research. Issues such as the export of genetic materials (Mudur, 1996: 464), unethical drug trials (Srinivas, 2004) and the commodification of genetic materials called for regulation. Two central bodies, the Indian Council of Medical Research (ICMR, 2000) and the Department of Biotechnology (DBT, 2002) formulated ethical guidelines and policies on the conduct of biomedical research on human subjects. With the rapid expansion of genomics-related research vis-à-vis the exposure of the diverse population groups that inhabit India, many issues surfaced on the handling, protection and use of genetic information. These issues led the central government to set up a national advisory committee. Apart from the complications related to its modalities, ownership, ethical, legal and social implications, the National Repository will have to deal with the structural and functional aspects of its relationship with source institutes (those who will share genetic resources and data with the National Repository). At the same time, the National Repository will have to take into account its relation with collaborative partners, including private pharmaceutical and bioinformatics companies vis-à-vis the source institutes.

With the initiative taken by the Anthropological Survey of India—a national institute under the Ministry of Culture—a 12-member committee headed by Dr S.S. Agarwal, the former Director of the Cancer Research Institute, Mumbai, was assigned to submit a report to the Union Government of India (*The Hindu*, 2006) by the end of 2006. The terms of reference of the committee were as follows: (1) to consider, evaluate and recommend the establishment of a permanent national repository for human genetic resources and data; (2) to suggest regulatory issues involved in such establishment at a national level; (3) to consider and recommend ethical, social and legal issues, including consent, intellectual property rights, benefits sharing, research, material transfer agreements, etc. of international standards, involved in collection and storage of DNA, cell lines and ancient skeletal material; (4) to suggest operational aspects of the repository, including policy issues; (5) to submit the complete report within six months of the date of constituting the committee (AnSI, 2006). The committee received many suggestions and technical inputs from members of civil society and from scientific communities. However, until now, the National Repository has not got off the ground. Explaining this, the former Director of the Anthropological Survey of India, Professor V.R. Rao, pointed out that 'it has been buried under the bureaucratic red-tape-ism' and 'many of the leading scientists are genuinely not wanting the National Repository to be formed because of their personal ego' (telephone interview, 4 August 2011). This view indicates that sharing biospecimens and data is not to the advantage of the careers and research of the collection owners. In response to a question on the desirability of a centralised national repository, the Director of the Centre for Cellular and Molecular Biology (CCMB), Hyderabad, Dr Lalji Singh (also a member of the advisory committee for the National Repository) opined that 'at CCMB we have our own biobanking facility. We do not need to join any other bank where we have to transfer or samples' (interview, 26 March 2007). Such a statement from a core member of the advisory committee does not bode well for the proposed repository.

Besides these bureaucratic and interpersonal factors, the planning of a national repository is pregnant with bioethics- and policy-oriented issues in biological research involving human subjects. It is evident that for the formation of a repository, the collection of genetic data is crucial. But what genetic data constitute and how they are generated or gathered in a particular societal context is of prime importance in the understanding of debates concerning bioethics in genomics. To understand the background of genetic sampling, fieldwork was conducted between December 2006 and May 2007 in three tribal and two caste communities living in four different states in India: the Dhodia tribe of Gujarat; the Bhil-Pawara tribe of Maharashtra; the Sahu-Teli caste of Chhatishgarh and the Kondh tribe; and the Agaria caste of Orissa. Interviews were held with researchers and doctors working at well-known public and private sector hospitals and research institutes in New Delhi, Hyderabad and Kolkata. The methods are described in Case 4 in Chapter 3 (p. 60). The interviews used in this case study were conducted in Hindi, English and other regional languages. The interviews with scientists, who are actively involved in genomic research and biobanking, and who are aware of the issues related to the establishment of the National Repository, indicate that the formation of the National Repository could stop the illegal supply of bio-samples to foreign laboratories, and avoid the accumulation of many separate biobanks of numerically small, vanishing tribes. Moreover, it could avoid repeat collection and study of samples from the same population groups, minimise costs and time spent on biobanking, and facilitate the central management of data. At the same time, fieldwork made clear that the sampling methods used currently and in the past may be a main obstruction to setting up the National Repository.

The deposition of samples into the National Repository will need to conform to formal bioethical standards if research and biomaterial exchanges are to be made feasible. A first issue to be dealt with is the notion of informed consent—both for prospective and retrospective data. In fact, researchers working in the areas of genomics and population genetics frequently consider informed consent more as a tool than as a bioethical principle, used in data collection among often illiterate participants to 'legally protect' the respective interests of researchers and participants (Patra and Sleeboom-Faulkner, 2007). Thus, when asked about informed consent procedures, Dr Munshi, a professor of physical anthropology at an Indian university heading a project on human genome diversity study on four tribal communities from central India, said

It is our longstanding relationship with the people that is enabling us to gather samples, not any assurance or incentive. Informed consent as it is understood in the bio-ethical and legal field, is neither feasible to take, nor is it effective. It is peoples' faith and goodwill in us, which is important...The procedure of individual consent is too risky and impractical in case of rural, tribal and illiterate societies. We receive their non-cooperation at the moment we ask for written consent. Usually, it is a matter of taking their signature or fingerprint on a piece of paper, declaring that they have understood the intent, the objective, the risks and benefits involved in the research for which they provide genetic data. This happens because they have had bad experience of property dispute or money lending associated with providing a signature and a fingerprint. Another issue is that it is morally problematic to ask for consent when the subject is not in a position to understand the inherent meaning of the issues and we ourselves are not sure about the future implication of the data (study 1, interview Dr Munshi; 12 April 2007; transl. from Hindi by Patra).

With regard to the process of data generation and informed consent, Mr Raji, a PhD research student jointly attached to a university department and a research institute that is working on the molecular aspects of genomic analysis, made the following comments:

> There is a set of guidelines that we were told to follow during the brainstorming session of the project's execution plan. But the real field situation is something different. You go to a tribal village where you do not get any kind of cooperation from the people who do not show any interest in your work...What we are given is research targets: to collect 50 or 100 samples in a week's time. It is unthinkable to establish the kind of rapport with the subjects that is seen when using typical anthropological fieldwork methods. Forget about educating and informing people about the risks and benefits involved, we use some tricks to get their blood samples. Our aim is to obtain the blood sample. Of course, we take maximum care not to misuse it. We normally take a medical team with us and declare in advance that a medical team will come to the camp near your village and will check your general health status, test your blood and distribute free medicine. According to the guidelines, we are required to take samples from unrelated individuals, but that is not possible in a camp situation. What we do is take blood samples from all individuals and collect the genealogy and other family details. And then we select the samples that we need for our purpose, and the rest we throw away. We know it is not ethically correct, but we do it, as it is very difficult to get people's support and their samples in less time (study 2, interview Mr Raji; 17 April 2007; transl. by Patra).

Owing to the difficulties associated with taking informed consent from illiterate people, the possibility of 'group consent' has been considered. Although 'group' consent in other contexts, such as in Canada among native populations (Tri-Council of Canada, 1997), is used as a safety valve against individuals providing samples against the group, 'group consent' can also be used to avoid difficulties in taking individual consent. In an Indian context, the concept of community consent is borrowed from traditional anthropological fieldwork techniques to achieve quick penetration of the community by obtaining approval for the recruitment of participants without antagonising individual participants. In practice, it neutralises the importance of individual informed consent procedures and hastens the data-gathering process. As one researcher opined:

> If we plan to take some samples from a community, we first approach the village or community headman, the local teacher, the local

representative or key person who can convince their own people, act as a buffer if there is any resentment, give some kind of protection to use and facilitate the smooth conduct of research. Once they give the permission, getting individual support becomes very easy. It is crucial in community research (interview Dr Patel; 23 September 2006; transl. from Hindi by Patra).

With regard to community consent, Dr MM said:

> In the generation of genetic data we just have informal consultation with the village/community leaders such as the village headman, the local schoolteacher or the local level Panchayat representatives. Though we give equal importance to individual consent, at times, with the approval and consent of community leaders, you can ignore or bypass the individuals (interview with Dr Munshi; 12 April 2007; transl. from Hindi by Patra).

Concepts such as 'community engagement' and 'community consultation' are ethical watchwords used in population-based studies of human genetic variation, where the theoretical aim is to allow human populations, that is, the subjects of research, to have a degree of meaningful control over the initiation and conduct of a study. The theoretical aim of community engagement or consultation is to enhance the population's control over the ways in which its members are studied. However, in practice, the notions of community engagement and community consultation become a strategic tool for the recruitment of individuals as subjects of genetic research.

It is clear, however, that the abuse of community and individual trust goes against the morale of ethics committees, who try to find ways of implementing ethical procedures. Dr Patel, a teacher of biological anthropology and one of the leading figures in genomic research, and also a member of numerous ethics committees in India, was more categorical about informed consent and other bioethical issues. He opined:

> We give maximum importance to informed consent in the data collection procedure. Written consent is the most preferred way; however, with the given population context, we do rely on oral consent with proper audio and video recordings. Now we are even thinking of using this method for all the subject participants. We even think of having the entire community consultation recorded…We take maximum care in maintaining the confidentiality of the subject

participants, even the name of the community we don't reveal to others and we do not even publish it. All the samples are coded and strict confidentiality is maintained in this regard...As a member of a national committee, I pleaded for the benefit to the participant subjects and communities. I argued for five percent of the research profit to go to the community and participants. But the detailed modalities are not discussed...I will be glad if you can show me an instance of benefit being shared with people from anywhere in the world. We in India have not reached that stage where benefit is seen in tangible terms (study 3, interview Dr Patel; 23 September 2006; transl. from Hindi by Patra).

With the kind of mechanism at work in India at ground level, where informed consent is not maintained properly and the identity of individuals is not automatically recorded, it is highly unlikely that justice can be ensured as far as benefit sharing is concerned. Although existing protocols (ICMR, 2000; DBT, 2002) discuss benefit sharing, the mode and mechanism of sharing between unequal players (powerful authority and less powerful subjects) is not fully discussed. Department of Biotechnology (DBT, 2002) guidelines say 'it is obligatory for national/international profit making entities to dedicate a percentage (e.g. 1%–3%) of their annual net profit arising out of the knowledge derived by use of the human genetic material, for the benefits of the community'. A prominent scientist and active member of many committees says:

> Though the issue of benefit [mostly financial gain out of research results] is in a conceptual stage, we need to talk and devise methods so that justice is guaranteed, but for that we need to have instances of benefit sharing in other countries. We in India have not reached that stage in genomic research in which benefits are tangible, so that a party can make a claim in the first place (interview with Dr Patel; 23 September 2006; transl. from Hindi by Patra).

At issue here is the question of what a benefit is and how participant individuals and communities view it in their own socio-cultural context. Another area of contention is how to determine who is a beneficiary. The issue of the ownership of the data (i.e. bio-samples) and the legal authority over the samples is problematic. Whether it is an individual, a family, a patient group or a community that holds the authority over the sample is a very tricky and complex issue. Paradoxically, the idea of sharing of benefit has created a poignant sense of moral conflict. Benefit sharing seems to sanction the commodification of human body parts, devaluing the worth of human life and failing to recognise human dignity.

The interview quotations provide an idea of the way in which informed consent is practised in India. It gives a glimpse of the gap between the official position based on ideals of bioethical protocols and the actual practice followed in the field. Dr Patel (study 1) represents the official position, whereas Dr Munshi (study 2) and Mr Raji (study 3) represent the field practice or situational dimension. These stated conflicting ways reaffirm the urgency of the question of whether the principle of informed and community consent is 'doable' and 'desirable' in the context of low- and middle-income societies, especially among tribal and small-scale communities. It also opens up a debate on the problems of standardisation of guidelines in data generation procedures and the effectiveness of institutional review boards in the streamlining and assessment of ethical clearances.

It is now clear that the idea of a permanent National Repository at a central level is not taking the shape initially envisaged by its committee. According to the views of many researchers, bioethical requirements stand in the way of it. But the bioethical issues involved in genomic research exist irrespective of the nature and size of the biobank, which may continue with or without the National Repository. The issues of informed consent, confidentiality, benefit sharing and public trust are hard to tackle. For instance, under informed consent, the issue of retrospective consent is potentially problematic in both the context of biobanking and in regard to India's decision to establish a central National Repository. The National Repository is considering the collection and storage of DNA samples from all anthropologically defined population groups in the country, and intends to gather them both prospectively and retrospectively from public, private and individual research centres. If the National Repository comes into existence it may need to obtain agreement from donors and sample providers for the transfer of their samples, in addition to acquiring permission for their future use. To trace the sample providers is a daunting task. Although the common use of blanket consent in India may make re-contacting donors unnecessary, if donors acquire the right to know about the new arrangements for the future use of their samples, the task may actually become impracticable. Additionally, public trust in the context of biobanking initiatives in India will be crucial to the success of biobanks or the National Repository, as there already exists distrust among the public about new technologies and their applications. With the increase in technological complexities, making efforts to stimulate public debate on the subject may be imperative.

Case 8: Transnational Biobanking and Institutional Dislocation in Indonesia[4]

Case 7 illustrated difficulties regarding genetic data acquisition when the collection and provision takes place in a situation of domestic financial, educational and social inequality. When inequalities involve a foreign scientist from a Western welfare society engaging in genetic sampling in a low-income country, no matter how noble or scientifically useful, its aims and intentions will be subjected to extra scrutiny. This case study on the biobanking activities of a Dutch scientist, Dr X, in the former Dutch colony of Indonesia illustrates how stereotypical views of biobanking practices in a low-income country portray it negatively as parasitical exploitation through hegemonic power, or positively as enlightening progress in a globalising world. A paradigm of progress regards biobanking in Indonesia as a scientific exploration of Indonesia's genetic make-up on a quest to identify disease and the cause of epidemics, while globalisation is seen as a treasure field of opportunities in which the local can be exploited and moulded through rational planning, shifting people and moving know-how. Entrepreneurs only need to take into account the cost of the movement of scarce resources such as capital, skills and sources of production to generate surplus value to the mutual advantage of partners in trade. Potential obstructions are usually identified as bureaucratic obstacles or as political plots against progress.

A paradigm of exploitation would regard Indonesia's underdevelopment as directly related to its past colonial subordination to the Netherlands, while its future seems to be dependent on lingering corrupt political systems determining the country's fate through its economic and political links with globalising hegemonic structures of power. Globalisation here means a capitalist ('Western') penetration of the local community and homogenising human diversity at the expense of local traditions. In this view, market mechanisms under capitalism and historical relations of socio-economic dependency have set the political stakes against poor, non-Western countries. Individuals as socio-economic beings here are victims of the capitalist system. In human genetic biobanking those individuals are objects of research (including genomics) and exploitative capital investment. As far as these individuals are allocated 'free and informed choice' to participate, individuals are treated as subjects with certain rights. But if these rights are not exercised as stipulated, they facilitate their exploitation as mere objects of research, and strengthen the position of entrepreneurs.

These two paradigms have conflicting views about how to advance the welfare and wellbeing of human beings. What they have in common, however, is that they ascribe subjectivities to individuals on the basis of the restrictive views of biomedical progress and exploitative biocapital. Underlying the paradigms of progress and exploitation are the notions of unitary capitalism and neoliberalism. But capitalism and neoliberalism are not eternal, homologous or without history (Zizek, 1991; Kipnis 2007); they are facets of our global political economy, including a moral one. The funds raised for biobanking ventures, the scientific knowledge gained in academic institutions and the access obtained to resources all derive from changing financial stakes, legal structures and institutional networks. And their room for manoeuvre depends on the space created by the political regimes and institutions within which biobanks operate.

An institutional understanding of global moral economies recognises that social, technical and economic institutions are interdependent, which necessitates a critical analysis of relations between the infrastructure of social, political and legal institutions of countries that biobanks tie together by developing their networks. But these networks are fallible. The subjectivities generated through the cross-fertilisation of institutions produce the condition in which networkers such as Dr X, introduced below, operate. For the quality of the atmosphere, trust, social hierarchies and connections, Dr X and other leading biobankers may capitalise on disjunctions between cross-national institutions and benefit from differences in national policies, standards of healthcare, bureaucratic efficacy, and public spending patterns on healthcare and sanitation. But some biobanking efforts fail. Although there may be financial and scientific reasons behind the failure, the examples below show how subjective factors in biobanking networks may be crucial.

As one of the eight Asian tigers, Indonesia started the 1990s with annual growth rates of 6–8%. But the unexpected 'Asian crisis' in 1997–98 brought decline: food prices shot up and the poor—by 1998 about 60% of the population lived below the poverty line—were hurt (Bergstrom, 2004). Recovery in Indonesia was comparatively slow, as it was complicated by problems accompanying the shift of power at the end of 32 years of authoritarian regime under President Sukarno. Although the following 'reformasi' was gradual and patchy, in 2002 the percentage of people living below the poverty line had decreased to 27%. Nevertheless, the GDP in 2002 averaged just US$994 per capita, the unemployment rate was 9.1% (Asian Development Outlook quoted in Bergstrom, 2004: 145) and the situation in the areas of sanitation, healthcare and nutrition was poor. Moreover, according to the United Nations Development Programme

Human Development Report 2001, 66% of the population was not using adequate sanitation and 24% of the population was not using improved water resources (in 1999). With regard to children under the age of five years in the period 1995–2000, the proportion of underweight children was 34%, the proportion of undernourished children was 6% and the proportion of under-height children was 42%.

Indonesia was not just waiting for high-tech solutions to its healthcare problems, although medical research was encouraged and its ethical regulation set up. According to a joint study by the Ministry of Health and the WHO, 26 sub-national ethical review committees were set up in 11 provinces. However, the study also showed that there was no single standard among these various committees with regard to membership and procedure, or the perception of bioethics in health research. It led to doubts about its effectiveness.[5] In particular, the capacity to oversee clinical research and compliance with guidelines are in need of reinforcement (Slamet and Elengold, 2002: 77). The inefficacy of research guidelines is partly related to a poorly functioning bureaucracy. Systemic corruption and a lack of legal certainty can also partly explain the uncompetitive environment for conducting business and slow economic recovery from the financial crisis in the late 1990s. Additionally, economic problems are due to recent acts of terrorism, unequal resource distribution among regions, a lack of reliable legal institutions, a weak banking system and endemic corruption. While traditional sources for financial and legal aid, health and social security weakened during the period of industrial and urban expansion in the 1980s and 1990s, governmental and non-governmental institutions have not been able to fill this gap (Abdullah and Hüsken, 2003). In short, without welfare institutions, effective supervisory organs and a financially secure population, families and individuals did not have the required support to decide about entrusting their DNA samples to biobanking projects.

On his quest to place Indonesia on the genetic world map, a Dutch scientist named Dr X set up the largest biomedical research centre in Indonesia using blood samples gathered from various laboratories, medical centres and the jungle. Invited by Sanbe Farma, a manufacturer of pharmaceuticals in Bandung (West Java), Dr X was to study genes for familial syndromes, such as cancers and forms of mental retardation, some of which are considered to be specific for certain ethnic groups in Indonesia. In 2001, Sanbe Farma announced its decision to start a repository of inherited diseases of the Indonesian people in all its genetic diversity. The project was advertised as 'a stimulus for future Central Government initiatives for a Human Genomics program to be established in this vast country of more than 220 million people and its enormous genetic resources' (cf. www.sanbe-farma.com/biotech.html, accessed 8 February 2006; new website available at: http://www.sanbe-farma.com, accessed 25 November 2013). Data used in this case study are based on insight into e-mail correspondence that took place between Dr X and a group of people that he involved in his plans to set up a HGDP in Indonesia. An anthropologist invited to participate in Dr X's project realised the relevance of this discussion to 'the exploitation of non-Western societies', and provided me with the correspondence.

In February 2002 Sanbe Farma was flourishing. At a global level, governments were pouring funds into genomics, and optimism reigned. Official institutional exchanges between Indonesia and the Netherlands in the area of biotechnology had started to increase in the 1990s. The visit of representatives of the KNAW (the Dutch Royal Academy of Science) to Bandung in February 2002,[6] led by Minister Hermans, resulted. The Netherlands had close links with the Eijkman Institute for Molecular Biology ranging from Indonesia and Singapore to the Netherlands. In an e-mail, Dr X relates how he had managed to befriend the Rector of the Bandung Institute of Technology (ITB), Kusmayanto, musing that he 'according to tradition will become the next Minister of Research & Technology', and proposed to start preparations for his new 'Medical Pharmaceutical Research Law' (21 February 2002). It was of great importance for Dr X to be treated with respect and, most importantly, to be accepted into the company's institutional culture:

> At work they have started to call me Pak Jo now, instead of Dr Jo, after the example of the Rector [Kusmayanto of ITB]. In short, I am accepted as one of them and no longer the odd man out. My young academics and other personnel worship me and it is a pleasure to work day and night (22 February 2002, to HJ; transl. by Sleeboom-Faulkner).

The emphasis here is on harmonious collaboration, hard work and confraternity. Dr X described working conditions at Sanbe Farma, which he found more satisfactory than those at his former laboratory at the Free University (VU, Amsterdam):

> And in fact I can do top research again without any further restrictions, and as well as in my laboratory at the VU: I no longer can see the difference between my laboratory here (over 10 permanent staff work there already) [which] has nearly doubled in size of what I had at the VU and with which I ended [my stay]. Last week, again,

I acquired a full-time molecular biologist (yes, promoted and all and top of the nation here) and a full-time pathologist. And one day soon we will purchase a Cytovision caryology system of US$100,000, while we are selecting the Flow Cytometer at the moment. You would not think it possible in Indonesia, would you? (22 February 2002, to HJ; transl. by Sleeboom-Faulkner).

Being in a low-income country, Dr X has to show that he can do world-class research there with first-class scientists and first-rate equipment.

When, in March 2001, the Genome Institute of Singapore was set up as the national flagship programme for genomic sciences in Singapore, Dr X saw an opportunity for sampling activities to traverse Indonesian borders:

In Singapore a 'Genomic Institute' has just been set up by a Prof. Dr Ed Liu,[7] who is pleased to coordinate things with me and with whom I want to observe all of the 500+ Southeast Asian ethnic groups by looking for support in the Philippines, Malaysia, Thailand and the rest of the 'region'. Of course, this is something else, something that even entire Europe cannot beat for all of these 'ethnic' groups are just 'large families' of course, who have lived in isolation in this gigantic archipelago for thousands and thousands of years (23 February 2002, to P en G (VU), B en F (KUN), D (EUR) (via HJ), C (RUG) en G (Ingeny in Goes); transl. by Sleeboom-Faulkner).

Dr X also needed to indicate that he had found genetic 'gold' not available to his fellow Europeans. Thus, Dr X estimated the chance of receiving Dutch governmental support as promising. Despite his widespread cynicism about the uprightness of Dutch political institutions, Dr X's faith in them was strong enough for him to believe that a reasonable proposal would be judged on its scientific merits. For this reason, Dr X expressed the hope that the Minister for Education, Louk Hermans, would 'play Father Christmas'. After all, Hermans had 'just given 2 million Euros to the charity for sports and children run by "Kus" [Kusmayanto, the Rector of ITB], surely leaving enough to spend some more—e.g., US$80 million before stepping down' (22 February 2002, to HJ). Through the Minister and the vice-director of the Board of Education and Science, Emil Broesterhuizen, he was introduced to Netherlands Genomics Initiative director Peter Folstar, who showed an interest in the idea of a joint Indonesian–Dutch genomics institute (23 February 2002). Although happy to have direct access to the 'highest Dutch genome', it did not diminish Dr X's distaste for following the 'bureaucratic procedures' necessary to gain access to funding:

> In the Netherlands I have developed an extreme distaste for writing proposals in vain, even though my score at the KWF was over 80% over 30 years of cancer research. So I have written a concise plan to Mr Broesterhuizen and I shall add a first draft of our planned home page to be called 'The Indonesian Hereditary Disease Registry' (22 February 2002, to PF; transl. by Sleeboom-Faulkner).

Dr X believed that his former success and position close to valuable genetic resources allowed him to disregard due procedure:

> Now I do hope that you won't be sending me a pile of forms, for then I will get up on my hind legs. I am so happy to have been able to avoid them here in Indonesia over the last three years. The precise contents of the plans will come later, me thinks [sic], and I shall welcome any suggestion from the Netherlands. We are prepared to take on anything as long as we can establish a beginning of collaboration. There are hundreds here raring to go but can't find a job and if they do they have no money to conduct experiments. A sad situation here, apart from with me at Sanbe Farma, fortunately (22 February 2002, to PF; transl. by Sleeboom-Faulkner).

Dr X saw his position fortified by his access to cheap educated personnel, whom, in his view, he was not exploiting, but giving an opportunity. Dr X seemed to know how to pull the strings in the Indonesian bureaucracy and understood how to approach the right person at the right institutions:

> Furthermore, we are setting up one or more 'field' teams that carry out expeditions, such as to Papua and the inlands of Sulawesi, Kalimantan and Sumatra. To approach small ethnic groups permits are required from a specially designated Ministry. All contact with the Government from here: NOT OTHER WAY ROUND PLEASE. So Louk, Emil and others, please, do not play the big guy in Jakarta. It is better if we, i.e., Pak Jahja, my boss at Sanbe[8] and 'Kus', the Rector of ITB, present ourselves as 'small boys' with an offer from the Netherlands. We know the 'right price' to pay (you don't, which would end in a right disaster) named KKN [corruption, collusion and nepotism][9] (try to translate that into proper Dutch) (23 February 2002; transl. by Sleeboom-Faulkner).

Thus, Dr X expected to pay officials to gain access to small ethnic groups, which he planned to sample 'in the jungle'. To both his pals and official contacts he boasted on his ability to manipulate the bureaucratic administration. Using cheap and motivated researchers in Indonesia, he argued that they would be ideal to conduct the sampling for him. And realising that healthcare service in Indonesia was problematic, Dr X planned to involve local healthcare organisations. Despite the fact that in Indonesia healthcare institutions can hardly cover the needs of the population, they were mobilised to cooperate in projects:

> Now that we have involved the Puskesmas [Community health centres] (68 in total) of Bandung, things are really starting to get cracking and the aim is to involve the local healthcare organisation of the entire country in an enormous national programme, for which there are more then enough means available (21 February 2002, to FH; transl. by Sleeboom-Faulkner).

Dr X, then, either had no clue about the existence of research regulation or he ignored it.[10] No mention was made of research ethics, such as informed consent and confidentiality of genetic data. Patient confidentiality would have been hard to maintain, as the health records were to be linked to the genetic identity of ethnic groups. On his website, he advertises, 'Requests for help in gene mutation searches will be given only if the pedigree is as complete as possible, if relevant medical records are sufficient'.[11] No effort is made to elaborate on how those who are found genetically susceptible to, for instance, familial cancer and mental retardation shall have to deal with their condition, and what precautions and what measures should be taken. These questions are the more urgent, considering that historically in Indonesian society there has been a tendency to regard patients with mental illness as a threat to those around them, rather than as sick people in need of support and care.

It was Dr X's institutional position that enabled him to mobilise contacts from Southeast Asia; apply for funding in the Netherlands, using his contacts in Dutch universities and government organs; employ cheap, highly educated labour in Indonesia; and pull strings in the Indonesian bureaucracy. Dr X's privileged position cannot be understood just as a natural result of the Dutch colonial and republican postcolonial eras. It was not so much the Dutch–Indonesian colonial history, but global developments in biotechnology that produced the institutional conditions for an entrepreneur-cum-scientist to mobilise potentially fruitful partnerships in biobanking. In the end, Dr X's efforts to set up a connection network did not even yield any funding, which was largely reserved for the Genome Initiative in the Netherlands.

Vis-à-vis his connections, Dr X put on a show of being respected far away from home in Indonesia, promising to move mountains at will. Both showing off how he rubbed shoulders with the world's greatest and giving the impression that he (and the other great) could take on the challenge of finding the key to Asian genetic diversity were not only an expression of his vanity, but, more importantly, conditioned by the institutions that he was in a position to mobilise and shape. What we saw was an attempt to mobilise 'cultural' capital to capitalise on the differences between academic institutions to advance the enclosure movement of the human genome.

But any judgement of Dr X as making use of the 'postcolonial' situation of Indonesia for exploitative purposes has to take into account that Dr X was Dutch. For others, such as Sangot Marzuki, behaved similarly: rubbing shoulders with people, using Indonesian scientists as cheap expertise, showing off their familiarity with Indonesia, receiving respect by dint of their foreign training, occupying a prestigious position, and using the local genetic diversity for biobanking purposes, while paying little attention to issues of social inequality and ethics. In 1992, Marzuki had been asked by former research and technology minister B.J. Habibie to lead the country's molecular biology research centre. On the condition that he would be leading an institute comparable to Singapore's Institute of Molecular and Cell Biology, he accepted. Marzuki came to Indonesia without any staff or laboratory, recruited his PhD students as scientists, bought equipment and invited his former PhD students from around Southeast Asia to help carry out research for the Eijkman Institute. Marzuki was given funds to pay six Australian scientists to work with him for three years, starting the AIMRI collaboration (the Australia/Indonesia Medical Research Initiative). The programme worked with 75 scientists in 1995, until the financial and political crisis in 1998, when the number of scientists had to be cut. With the change in government at that time, future technology was no longer a priority (Sabarini, 2010).

The institute continued work on genetic diversity, the field that Dr X had proposed to develop. Marzuki and colleagues collected more than 1500 samples at the Hasunuddin University Medical School in Kakassar and at two high schools—one in Pangkajene and one in Takalar. They chose 28 major populations of the Indonesian archipelago. The extensive questionnaire covered questions on marriage patterns and recorded common diseases, such as malaria, hepatitis, diabetes mellitus, preeclampsia and kidney problems, and signs of mitochondrial disorders, such as blindness, deafness and neurological symptoms (Marzuki et al., 2003: 6–9). Though aware that ethnic strife and population and migration politics are particularly sensitive in Indonesia, the authors did not specify ethical and social issues. The authors only say:

> In all cases, blood-sampling expeditions were carried out with the support of the local medical scientists and health officers. This overcame any language barrier and helped in obtaining proper informed consent and in other ethic, legal, and social issues (2003: 8).

Preparing the way for Indonesia's human genetic repository, Marzuki believes Indonesia's research and development capability will improve over time through international collaboration in the field of research, especially with America.

Observing the biobanking efforts of Dr X and Marzuki, it becomes clear that the paradigms of exploitation and progress are too simplistic to understand their respective developments of transnational collaboration in the life sciences. Dr X's venture into biobanking in Indonesia cannot be understood adequately through the postcolonial 'exploitation' paradigm, even though it is clear that Dr X was driven by paternalistic enlightenment ideals. Viewing Dr X as exploiting native genetic treasures misses the point that Dr X had been invited to do this by Sanbe Farma and that Marzuki, who had been in Australia for over 25 years, had been invited to do just the same in the Eijkman Institute. In this instance, the 'exploitation' paradigm is a result of a nativist form of anti-Westernism, the critical potential of which to analyse the ethical and socio-political issues at stake is reduced when it comes to similar activities engaged in by countrymen. In both cases the contact networks, social hierarchies, institutional collaboration and the shared subjectivities of what is prestigious and important research were crucial to the continuation of their biobanking efforts, and both cases capitalised on disjunctions between cross-national institutions. When analysing discourses on modern public health and biomedicine there always seems to be a positive side, which sheds light on the desirable benefits of health and modernity, including hygiene and medicine, and a negative side, which regards public health and biomedicine as a mode of social control, a coercive force and unethical enterprise, including human experimentation and eugenic practices (cf. Charkrabarty 1997; Rogaski 2010: 156). The ambiguity of the biobanking efforts discussed here points at an ideological entanglement of both paradigms. And its outcome will not just depend on the institutional and educational infrastructure of its practice and the objective, material situation of the sampling population, but *also* on the subjectivities of biobankers, genetic samplers, the population at large, and the medial and social critics.

Case 9: The Taizhou Biobank: Identity and Bioethical Capital[12]

Case 9 analyses national efforts to set up a biobank and the acquisition of the bioethical jargon accompanying it. Bioethics in Chinese biobanking was initially institutionalised to prevent 'genetic theft' and to facilitate the advancement of the modern life sciences. Its expected functions of protecting China's genetic property and stimulating China's development in the life sciences have strongly influenced the ways in which bioethics capacity is being built and institutionalised in China, including the role of individual choice and responsibility. As the ethical aspects of the research were especially geared to protect national stakes, the interests of individuals received less attention. It remains to be seen how bioethical concerns will play out in the political context of China's economic growth orientation, its strong belief in science as a vehicle of progress and its nationalist discourses.

By dint of the increased use of biobanks in China, the Chinese state and scientists have made great efforts to establish bioethical guidelines, including those for approaching individuals and families for their informed consent to medical procedures and for requesting individuals' informed consent before the 'donation' of tissue samples to store in biobanks. Until the 1990s, dominant Chinese discourses of development described China's position in the global environment in the language of nationalism and international realism. And after the economic reforms and China's Opening Up in the late 1970s, international relations experts regarded selfish competition as the driver of world change (Wang, 1995; Sleeboom-Faulkner, 2004). But since the late 1990s China has started to describe itself as both in competition and in collaboration with what it regards as societies advanced in the life sciences. In the past, collaboration was regarded as an expedient for China as a developing country, which had been 'isolated from the world' for decades, and now had 'to catch up with the advanced world'. In the new global outlook, the 'development' paradigm in the life sciences is being replaced by that of 'collaboration', in which China presents itself as an equal partner with similar goals and sharing the ideal of mutual benefit. This change of direction had major consequences for the ways in which it has come to formulate bioethics.

Although domestic genetic sampling activities precede the 1990s, it was only in the second half of the 1990s that the genetic sampling of poor individuals in remote mountain areas became problematic in the context of their mass export to the USA (Pomfret and Nelson, 2000; Sleeboom, 2005; Guo, 2009). An international diplomatic crisis erupted when an American Chinese man from Harvard University, Xiping Xu, under contract with a company called Millennium, renewed his connections with his alma mater in Anhui Province and started to organise mass sampling activities, the results of which were exported to the USA. After the *Washington Post* revealed this story on 'genetic theft', international scandal resulted. Although there was outrage about the possible harm done to poor Chinese in isolated regions, the main uproar concerned the fact that 'China's rich genetic resources' were being exploited and robbed by Americans (Xiong and Wang, 2001; Yang, 2003; Guo, 2009). Immediately after the scandal similar international collaborations were cancelled, and interim legislation was produced for the import and export of biomaterials, which came into effect in 1998. At the time, however, no discussions were held about the use of bioethics regulation in the context of China's Human Diversity Project—the Kunming Ethnic Biobank—and other bioethically-charged issues related to the banking, preservation and research of human genetic material. Although the genetic sampling and flow of biomaterials across the border was protected and supervised through strict guidelines, bioethical issues of genetic sampling activities within China did not appear in the dominant press. When such discrepancy was pointed out to officials and scientists, the usual response in the 1990s and in the beginning of this century's first decade was that China, as a 'developing' country, did not yet have suitable bioethics institutions to deal with these issues, but was in the process of setting them up. The self-image of China as a developing country in this case study is complicated partly by the discursive value of development as a crucial stage towards an ideal society (see also Yan, 2003; Pun, 2003). In such discursive context, being 'only' a developing country had strategic value in the process of building the life sciences, as it can explain regulatory loopholes; and as a nationalist quest, the discourse of development contained a strong element of positivistic belief in the imminent rise of a strong China.

The example of the Taizhou longitudinal cohort study illustrates the novelty of a situation in China in which bioethical requirements take account of individual informed consent in the context of a collective undertaking aimed at strengthening China's competitiveness in the life sciences. I argue here that, even though international collaboration is instrumental in setting up the Taizhou biobank, the development of bioethical capital (Franklin, 2003) leans heavily on older processes of international, national and collective identity formation when mobilising individuals as donors. The Taizhou biobank is earmarked as China's future national biobank, and is expected to become gargantuan in proportion.[13]

Taizhou, in Jiangsu Province, harbours what has been called the world's largest human genetic biobank in Taizhou's Medical Hi-Tech Industrial Park. China Medical City (CMC), a quasi-governmental organisation established in September 2006, hoped that the genome data bank would help the city leap ahead to the cutting edge of twenty-first-century medical developments. According to Wang Jingsu, the CMC's deputy director, the medical park needed to invest an estimated US$10 billion in its first five years, 20–25% of which it would obtain through central, provincial and local government funding.[14] CMC hoped to attract the rest of the required funding through foreign investment. Mr Wang claimed that CMC collected samples from people on a voluntary basis, and it had obtained permission from housing committees in different areas within Taizhou to visit residents to ask for their participation (Toland, 2007). Considering that it had a five-year plan to collect genetic data from one-fifth of its five million citizens, it was puzzling to Europeans that the biobank expected such a high response rate (Palmer, 2007). Furthermore, neighbouring Japan has had extreme difficulties finding volunteers for its national cell bank (Masui, 2009), and in Taiwan the entire venture of a biobank is considered to be a failure precisely because of the ethical issues involved in donor recruitment, in particular that of informed consent (Yang, 2007).

But after the BBC broadcasted bioethical concerns related to the Taizhou Biobank in 2007, the managers of the biobank radically adjusted the narrative of the bank. In the article 'Rationales, Design and Recruitment of the Taizhou Longitudinal Study' (Wang et al., 2009), the Taiszhou biobank was transformed into the 'Taizhou longitudinal study' with the aim of finding cures for increasingly prevalent non-infectious lung and heart diseases, and cancers. These disorders, it was argued, now prevail in China just as in other modern countries and unlike low- and middle-income countries (Wang et al., 2009). The former target of many millions of samples is no longer mentioned, and a more modest base line survey target of 100,000 people is indicated for completion in two years' time. The bankers published an article in *Biomedical Health*, scientifically justifying the research by arguing that this human genetic cohort study would expose many environmental and genetic correlates important to the aetiology of the targeted diseases. The study made use of a roster of persons aged 30–80 years old, obtained from the offices of the Public Security Bureau, Bureau of Statistics and the Community Committee. It showed how bioethical procedures were in place, including the house-to-house distribution of materials, personal interviews by expert interviewers who knew the Taizhou dialect, the taking of informed consent, and the anonymisation of collected data pertaining to socio-economic status, personal behaviour, family medical history, reproductive history, psychological and physical examination results to protect the privacy of the interviewees. This information and data on mucosal and blood samples were stored electronically. While follow-up studies are carried out every three years, study participants will also be followed-up indefinitely by using a chronic disease register system for morbidity and case-specific mortality already established in Taizhou community-based primary health centres. At the same time, the managers of this study claim that the samples are entirely de-linked from the cohort members. The aim of conducting follow-up studies as described in the article raises methodological questions concerning the access of the data of cohort members.

An important juridical and financial issue in biobanking is the marketisation of knowledge resulting from research, and the decision of how to share the benefits yielded from the study. It is clear that the hospitals providing information and samples to the banks will benefit to some extent, but it does not speak for itself that the cohort members or their families will benefit. A well-known bioethical issue in biobanking is whether it is ethical when participants gain little from their 'gift', while researchers are allowed, via patenting law, to obtain financial profit from their research findings (Toland, 2007; Dickenson, 2007; Winickoff, 2008). In the case of countries without adequate healthcare provision, such as the People's Republic of China (PRC), an important issue is who will benefit from the knowledge and medicine the biobanks are supposed to yield. Most biobanks in Europe, such as the UK Biobank, operate on a public health model in which little or no personal health feedback is offered, donations of human tissue are conceptualised as gifts in service of the public good. This notion of public good rests on the Lockean idea of a social contract between donor and banker: the donor gives in the expectation that, ultimately, others in need of healthcare will receive (Dickenson, 2007). To rely on assumed altruism and notions of the public gift, while focusing on the future commercialisation of research results, seems to deny the reciprocal principle, and is made possible only in the context of the high expectations of global medicine and life science, particular Euro–American notions of bioethical capital and particular Chinese notions of collectiveness.

In fact, China, although having abolished the idea of central funding for collective healthcare in the late 1970s, has kept up its emphasis of socialist service provision in the healthcare sector. Thus, PRC healthcare provision is formally based on a socialist notion of health provision, claiming that hospitals and physicians work at the service of the people (Yang, 2008). Nevertheless, the system is run on free-market principles, and the majority of the population has no healthcare insurance (Bloom et al., 2008). Nevertheless, it is clear that the mass media [or 'propaganda' (*xuanchuan yulun*) in Chinese] can persuade people to help the Greater Common Good when convinced this is humane, socialist, Chinese or the right thing to do. Thus, blood donation is relatively successful in urban areas (Adams et al., 2009). Moreover, continuous media messages expressing a belief in the fruits of scientific progress and national advancement in the life sciences, and the frequent official announcements of healthcare reform arouses a sense of expectation about the Taizhou longitudinal study, augmented by enthusiastic students whose expertise in interviewing expresses ethicality and thoughtfulness. Moreover, the project is supported by the state government, the city council and the hospitals, who all sing the same optimistic tune of progress, science, 'service to the people', Chinese socialism and bioethics.

Although many issues could be raised concerning bioethical aspects of the Taizhou biobank, Case 9 has focused on the way in which a global moral economy of bioethics was adopted. This biobanking project flexibly adopted and accorded its own meaning in the application of bioethics. Couched in the official socialist jargon and political lingua franca of science enlightenment, national progress and global bioethics, the example shows how the language of bioethics was mandatory in keeping up the international image of the bank, which is crucial to its current and planned international collaborations. The example also showed how the rapid adoption of the promissory language of biomedicine, bioethics and informed consent coexist with the current imbalanced access to healthcare, a politically correct atmosphere of informed consent and a 'socialist' spirit of healthcare service for the masses.

Conclusion

Case 7, on the government initiative to set up the Permanent National Repository for Human Genetic Resources and Data of India, indicates that, in India, there exists a demand for a large databank, which needs to be balanced against the local interests of owners of repositories and the difficulties associated with international standards for regulation. As at least some local repositories prefer to continue to work with their present collections to sharing these with other researchers, there is an indication that local competition is strong. But even those prepared to share their sample collections find the regulatory requirements of a national repository that needs to comply to international standards problematic owing to the bureaucracy and extra work involved. Crucially, newly adopted bioethical guidelines may reveal that sample-taking of existing collections does not conform to ethical requirements, thereby making their use unacceptable. These problems also play a role in Indonesia and in the PRC, although they are played out differently. Nevertheless, the cases presented in this chapter aimed to highlight different aspects of biobanking in light of particular assemblages, so that there is no space to discuss these problems here.

Case 8, on the biobanking initiatives in Indonesia, illustrates some of the subjectivities involved in the acceptability of biobanking efforts in light of paradigms informed by postcolonial and modernisation theories of progress. By describing practices and activities by two biobankers in Indonesia, Case 8 indicates that neither ideas about 'imperialist exploitation' nor 'medical progress for the people' are very helpful in understanding biobanking activities. The respective biobanking initiatives led by Dr X, at the invitation of Sanbe Farma and by Marzuki, at the invitation of Minister Habibie, show that Dr X and Marzuki had similar aims, used similar methods and had similar attitudes to scientific results, banking and ethics. Nevertheless, critical theories easily dismiss 'European' biobankers as vehicles of exploitation, and 'native' biobankers as bringers of progress. Only a closer analysis of the life assemblages involved can reveal differences in the biobanking practices of relevance to their functioning according to local needs.

Case 9, on the Taizhou longitudinal cohort study in the PRC, shed light on the ways in which issues pertinent to the use and ethics of biobanking efforts are presented to the public, and how biobankers whose thoughts are immersed in nation-state building adopt alien frameworks of thinking directed at the people the biobanking efforts claim to serve. The case study raises the question of to what extent local practices of sampling and banking samples follow the alteration in protocols indicated by the change of discourse, an issue that only empirical studies on biobanking efforts can shed light on.

Only by situating biobanking activities in their global contexts can we appreciate the transactional spaces and transnational traffic involved in biobanking activities, which, in turn, need to be understood in the life assemblages they are part of. The transnational perspective on biobanking used in this chapter shows that the moral pillar of informed consent, instead of empowering, can be disempowering, as its meaning is easily misread, corrupted or ignored in the life assemblage it is part of. Nevertheless, informed consent procedures are necessary for biobanking activities that need international recognition. Furthermore, in the long run, when the public becomes familiar with foreign notions of informed consent, a sense of mistrust may grow among the public, especially when biobank advertising claims to work for the benefit of humankind, and contribute to global medicine, national development and science, while doing lucrative business.

Cohort members may not be aware of the possible consequences of storing body tissues and sample and lifestyle information about individuals, and may not think of the importance of discussing sensitive issues about privacy, trust and the question of who will benefit from biobanks for many reasons. Donors that do not have the ability to nurture, sustain and develop themselves may not care about personal rights or informed consent when they are asked to donate under pressure. Concepts of personal and communal rights seem to be misplaced when potential donors are not in a position to worry about the long-term consequences of the consent. If biobanking is to benefit local communities and societies, we need to think about how the donation of samples can benefit the community in tune with its corresponding life assemblage. The consideration of local life assemblages would have implications for, first, the question of whether donation for immediate payment should be practised; second, the questions of who defines the purpose of biobanks and how the research they enable is defined; and, third, the way in which intellectual property rights are organised when built on commercial biobanking. This chapter, in short, shows how the moral economies on a global, national, local and corporate level interact in the context of various life assemblages.

5
Life Assemblages of Human Embryonic Stem Cell Research in China and Japan: Bioethical Problematisations and Bioethical Boundary-making

Views about bioethics and the way it has developed and spread over the globe are manifold. But both patrons and critics of neoliberalism assume that 'Western bioethics' have shaped and dominated the formation of bioethics in the world. Neoliberal views that subscribe to the notion of a universal bioethics doubt that indigenous views in some cultures respect the privacy and autonomy of the individual. Training and collaboration would persuade other cultures of the fairness or at least the expediency of the 'universal' bioethical views developed in the so-called West to the current international politico-economic order. Bioethics, in this view, would provide a degree of 'social justice' and increase the choices available to individuals. Critics of neoliberalism argue that what they regard as hegemonic 'Western' bioethics prescribes guidelines and regulation to other cultures under the economic pressures of capitalism and through international hegemonic neoliberal institutions. In this view, the assertion of 'Western' bioethics institutions enables forms of life science research that would be dismissed as exploitative and irresponsible without these institutions. Not adopting 'Western' bioethics, would simply discount other cultures as uncivilised or 'lacking human rights'. There is a third view common to intellectuals and activist movements at home, which supports the bioethics of the 'native'. 'Native bioethics' usually opposes 'Western bioethics', claiming that bioethics and so-called human rights are not universal, and that indigenous views on bioethics should replace those of 'the West'. Such bioethics would be 'fair', based on local cultural norms and values, and serve the interest of 'the people'. Such 'native bioethics' usually claims to represent 'domestic culture' and often propagates a generalised stance that pits 'native views' against 'Western views' (or 'neoliberal views'). In some countries, however, such 'native views' represent dominant domestic stakeholder views, which can potentially violate the wishes of minorities. This chapter will pay attention to all three views in the context of culturally sensitive debates around human embryonic stem cell research (hESR) and human cloning. The three general stances simplify the multitude of views in reality, but are here used to explore problems in discourses belonging to particular life assemblages.

Largely drawing on empirical field studies in China and Japan, this chapter delineates new moral economies that are developing around hESR in relation to global dominant bioethical views. These new moral economies belong to particular life assemblages that incorporate a constellation of stances that can be typified as universalist (neoliberal and anti-neoliberal) and nativist, but which are expressive of the problems inherent to the particular histories of life assemblages. Universalist stances claim that international collaborations in hESR serve the dissemination of a universal body of bioethical guidelines. But the discussions in case studies in this chapter show that, first, it is questionable whether this 'trickle-down effect' occurs in practice; second, the cases indicate that the view that bioethics is merely a result of capitalist pressures and neoliberal ideologies of 'hyping' (Sunder Rajan, 2006: Ch. 3) is not justified; and, last, that the nativist forces aiming to shape an Asian bioethics are hardly united. This chapter uses the notions of bioethical problematisation, and bioethical boundary-making to show how global, national and local factors are interwoven in shaping bioethics discourses and practices of bioethics in human reproductive cloning and hESR.

The cases in this chapter focus especially on human reproductive cloning and hESR. The case study of human reproductive cloning is relevant to discussions of the global moral economy of health because a worldwide rejection of human cloning is presumed to exist. But, in practice, such rejection takes idiosyncratic local forms shaped by public policies on and cultures of health and science. The case study of hESR sheds light on how health interests are defined variously in different political setups, and whose health interests are accommodated, when and to what extent. For the investment into potential stem cell cures for a disease has to be weighed against the cost, harm and interests brought about by the disease. The responsibility involved in finding cures and the research ethics involved have to be understood in the discursive context of the 'bioethical problematisation', and as part of the healthcare and research practices of a life assemblage.

The theoretical notion of bioethical problematisation refers to the way in which bioethics is constructed as problematic, the norms and values used in problem definitions, and the action plans recommended as acceptable solutions, including an assessment of 'ethical' costs and benefits. Bioethical problematisations are powerful in terms of moral economies of health because they do not simply reflect a reality that exists either in the past or in the present. For they may actively create a new reality by shaping what is thinkable in the domain of life and death, and remake the moral world in which scientists work and 'the people' live in. It is especially scientists, policy-makers and contributors to the bioethical literature who are involved in creating such new worlds in which scientific possibilities are translated into bioethical acceptability. Thomas Gieryn's concept of boundary-making (1983, 1999) showed how, through boundary work, scientists try to expand their authority to other knowledge fields, monopolise professional authority and protect their autonomy over professional activities. In his article 'Boundary-work and the Demarcation of Science from Non-Science' (1983), Gieryn speculated on the strategic value of the reputation of non-scientific thinking to scientists. Apart from the distinction between science and pseudoscience, it will become clear that the distinction between ethical and non-ethical is important in the boundary-making in the life sciences, and belong to a particular life assemblage.

   Although for some time bioethical scruples regarding stem cell research—in particular hESR, associated with the destruction of human embryos or human life—has been seen as the domain of wealthy welfare societies (Sleeboom-Faulkner and Patra 2008), it is clear that bioethical science has also become important in the valuation of what is regarded as 'good science' in Asian countries. The claim of performing 'good science' is also highly charged in China, where regulatory authority is disputed, funding resources are relatively scarce, and a political connection and an international academic network are regarded as crucial to the continuity of research. The same is true, however, in Japan, where bioethical mis-behaviour is generally thought to lead to scandal, and with it the with-drawal of support by colleagues, funding agencies, patients and clients. The introduction of bioethical regulation in the context of hESR involves much more than the safety of donors, the protection of researchers and the legitimisation of research. In fact, the myriad of often entwined factors of the life assemblage play a role in scientists' need to produce what is regarded as 'good science', including the acquisition of government funding, which is increasingly linked to the requirement of bioethical review, publication in international peer-reviewed journals, the facilitation of international science collaborations and the increasing need for a good image, both internationally and at home.

On the basis of a study of the bioethical problematisation of human reproductive cloning in China, Case 10 falsifies preconceived ideas about the ways in which bioethics develops in Asia, be it in resistance to 'Western imperialism', following neoliberal ethics, or through the creation of native bioethics. Case 11, on discourses pertaining to the regulation of therapeutic cloning in China and Japan, indicates that bioethics in embryo research correlates not so much with the moral status of the embryo as with the way morality can accommodate pressures of dominant actors. It will be shown that the formation of a bioethical problematisation of therapeutic cloning depends on the ways in which the pressures from those directly involved in embryo donation (donors, researchers, medical professionals) are connected with and accommodated by policy aims of scientific progress and innovation, and concerns of population control. Case 12, using the notion of bioethical boundary-making, shows that Chinese and Japanese science discourses around the ethics of hESR are shaped differently, and how the discursive relations between relevant voices in the discussion are shaped predominantly through external (international) and internal (domestic) pressures in the context of their own respective life assemblages.

Case 10: Human Reproductive Cloning in China[1]

The claim that, given the availability of the technological and scientific capacity, nothing can stop scientists from developing any, including unwelcome, applications expresses both a sense of fear and fatalism. The worry behind this claim is that there are always opportunities for amoral scientists, such as Dr Moreau or Dr Frankenstein, and terrorists to develop dangerous technological applications somewhere in a hidden backroom or on a deserted island far away. Although not impossible, advancing such secretive research may not be easy without access to advanced equipment, years of training and colleagues contributing to the desired technological aim. Human cloning is an example of an application that, though condemned worldwide, arouses fears of its stealth realisation. Having been discussed globally from many angles and with great concern (Bonnicksen, 2002; Holland et al., 2004), international agreement has now more or less dictated the prohibition of human cloning. This, of course, does not mean that everyone in the world opposes human cloning (Savulescu, 2009; Harris, 2011), and it does not mean that all countries reject human cloning for the same reasons. China, for instance, had its own reasons for opposing the 18 February 2005 United Nations (UN) motion that aimed to forbid human cloning entirely. Although some may argue that there are 'Western' forces that have pressurised Asian countries into adopting an anti-cloning stance, it is very clear that Asian countries have not imported 'Western' views on this topic lock, stock and barrel. Debate on human cloning in China, for instance, has developed following its own dynamic, of which traces can be found in novels, the media and in educational materials. The debate can also be found in the curriculum of life science students in the form of textbooks for medical and bioethics. An examination of these textbooks provides insight into the considerations important to medical and bioethical problematisation of human cloning, and it shows the way bioethical research regulation takes shape and for what reasons. Such insight is important to issues related to whether and how regulation on the subject is implemented in practice.

To understand the regulation of the life sciences in China, it is important to know the significance of state intervention in that country. State interference with science in China in the form of regulation or otherwise has a particular sensitive history. According to contemporary official views in the People's Republic of China (PRC), science should be free from ideological and political interference. This view carries poignant meaning considering that in the first decades of the PRC, from the 1950s, the biological ideologies of Lysenko and Michurin, imported from the Soviet Union, damaged the field of genetics and the careers of many scientists. It has led to a general self-understanding of life scientists as international scientists and an aversion to external interference with research, especially based on state ideologies. In this spirit, when, in 1978, science became one of the Four Modernizations under Deng Xiaoping's reign of reform, biology was to receive protection from political intrusion. And in the 1980s, the life sciences received ample state support and became central to national science programmes in the 1990s, including that of hESR and therapeutic cloning (Sleeboom-Faulkner, 2008a).

Textbooks on medical and bioethics provide a wealth of information on the kinds of discussions, criticism and ideas around human cloning and modes of regulation in China. The method I used in this study involved various steps. I collected a few dozen of the textbooks, which are for sale in ordinary bookshops, over the last decade. Chinese academic books, including these textbooks, usually have a detailed table of contents, often covering over three pages of headings. These tables of contents clearly show the structure of the book and are an indication of the categorisation of cloning and stem cell research, if covered. I studied the time of appearance, the categorisation of, and the space allocated to, the topics in all of the 17 textbooks used. Further, I examined why attention was paid to the topics and how they were represented, while focusing on the way the textbook authors delineated 'Chinese' discussions from 'Western' ones, and how official guidelines regarding cloning and stem cell research are represented and embedded in the textbooks. As the textbooks are written in Chinese, in cases of doubt about the correct translation or in cases of ambivalence, I have provided the Chinese in pinyin, the mainland Chinese alphabetical transcription of Chinese characters. In the text, when I refer to the medical textbooks, I use the numbers that I have allocated to them in chronological order.[2]

In view of its troubled history of political meddling with science, any constraint of science has been viewed with trepidation and suspicion. Authors of textbooks in the PRC therefore have had to tread carefully when justifying such interference. The textbooks are produced by academic specialist in bioethics, philosophy and medicine, published by academic publishers and directed mainly at an audience of medical students. The writing and publishing of the textbooks largely reflects the officially recognised need for discussion on bioethical issues and the perceived need to provide medical doctors and researchers with answers to moral dilemmas. The textbooks themselves say that they provide moral guidance to future physicians and researchers that otherwise may lead to scandal, dissatisfaction and official embarrassment. But the textbooks are also a result of a structural tendency in official academic institutions to appropriate a field and define it in the terms of currently important issues in political debate, varying from socialism and liberalism to Confucianism and nationalist expansionism. An examination of 17 textbooks on medical ethics and bioethics sheds some light on understandings of the relation between political ideology, bioethics and science presented to students of medicine.

Textbooks on medical and bioethics aim to provide students and researchers with an overview of issues important in medical ethics, science regulation and discussions about current medical problems. The textbooks claim to describe a new objective scientific discipline related to other social science disciplines, not subjective moral guidelines. However, in the dialectics of traditional Marxist jargon, objectivity is not disturbed by class bias but actually based on it as a reflection of changing class relations. Over ten years, this new discipline has undergone various changes, and the textbooks reflect this. Apart from the eras recognised in Marxist stage theory as determinants of biomedical morality (primitive society, slavery, feudal society, capitalism and socialism), tradition has become recognised as a determining factor of what is seen as biomedical ethics. Although biomedical ethics is regarded as part of the superstructure of society, also referred to as the realm of 'spiritual civilisation', it is viewed as an important tool for guiding scientific development and its applications in the medical field. It is the task of intellectuals, then, to guide the people's biomedical ethics in a direction suitable to a country in the first stage of socialism with its particular ancient tradition. Thus, the ethics of cloning and hESR discussed in the textbooks is formulated and justified with the policy of establishing socialism with Chinese characteristics in mind.

While in the mid-1990s hardly any attention was paid to the issue of human cloning, by 2005 entire sections of textbooks on medical and bioethics were dedicated to it. Human cloning technology is regarded as a key technology in biological engineering, with potentially wide application in the life sciences and pharmacogenomic medicine. The applications mentioned most frequently are the generation of knowledge on the genetic and biological development of human life, human reproduction, and the diagnosis and development of medicine and therapies for serious diseases, such as heart and lung diseases, and diseases of the nervous system. But only after 2003 was the theme of human cloning split into the two sub-themes of reproductive and therapeutic cloning. As in many other countries, this subdivision was to have great implications for the development of science regulation. Reproductive cloning was to be associated with an 'unethical' form of reproducing 'identical' humans, while therapeutic cloning was to lead to a cure patients using a technique called somatic cell nuclear transfer, which has more positive connotations in countries that allowed its development. In most countries, human reproductive cloning was prohibited by law, although most also have proponents of the technology.

Arguments for and against human cloning occurred in 15 out of the 17 medical textbooks. A definitive ethical stance is taken on human cloning, which reflects government policies on the subject. Thus, on 19 March 1997, health minister Chen Minzhang announced China's view on human reproductive cloning: 'not consent, not support, not allow, not accept' [Bu zancheng, bu zhichi, bu yunxu, bu jieshou]. But the PRC has not produced legislation against human cloning in general. For China preferred to closely monitor research on embryo cloning, *prohibiting human reproductive cloning only*. By allowing therapeutic cloning, the argument went, 'research clones' can be used for the production of medicine, the cultivation of organs and the advancement of scientific insights (9: 213–14). Concerns most frequently mentioned in relation to human reproductive cloning are related to the views people have of how society should be ideally organised and developed. Consider the arguments consistently used over time in favour of the use of human reproductive cloning:

- to improve the stock of the people (preserve the best human genes; maintain a high quality gene pool) (6; 10; 12; 13);
- to create superior human beings (mentioned are, e.g., Einsteins and Newtons, sportsmen, heroes and scientists) (4; 10; 11; 13; 14);
- to help infertile couples have offspring (4; 6; 11; 12; 14);
- to control the development of human cloning, as no regulation can stop it (14).

The fact that the authors list these arguments does not mean that they support them. On the contrary, the textbooks list all the arguments they believe are important to the debate in China, both for and against. Of these counter-arguments, I discuss five that are emphasised in the textbooks for reasons related to conditions in Chinese society.

First, human reproductive cloning could damage human evolution (6; 8–10; 12; 13; 16; 17). A concern with the human ability to steer evolution in the right direction is evident in the list of arguments for and against. But if it were possible to control the genetic development of human evolution, the textbooks discussing the topic have no doubt about its political realisation. This view expresses the belief in the power of the government to set goals for the development of the Chinese people.

The view is implicit in a second, related counter-argument: owing to the, for as yet, inability to control human evolution, cloning technology could augment the currently lopsided population growth (2; 4; 6; 10; 11; 13; 17). Thus, Zeng Jianmin provides the argument that in feudal, backward countries reproductive cloning could seriously disturb the gender ratio (4: 154). Textbook 17 is concerned about plans for eugenic cloning, and asks rhetorically who will decide which human traits are to be cloned? If left to individual governments, cloning would surely be used in biological weapons, with serious implications for society. It would be potentially more disastrous than nuclear weapons, allowing the large-scale cloning of murderous criminals, the cloning of human primates, human–pig and human–horse hybrids. If left to the couples themselves, the textbook asks rhetorically which partner we suppose will be cloned (implying males in heterogeneous marriage) (17: 175–179). Most discussed in the textbooks is the third counter-argument, that reproductive cloning would damage the current notion of the family and the family household (2; 6; 8–12; 17). It is feared that if reproductive cloning is allowed, forms of family organisation and households would multiply uncontrollably. Moreover, by separating reproduction of the population from relations of intimacy (qinggan) the drive to form a family would lose its force (e.g. 2; 10; 16). Men would no longer be needed (e.g. 6; 16) and, potentially, five people could become parent to one child (semen donor, egg donor, surrogate carrier and social parents) (6; 10). According to textbook 8, without a consanguineous relationship, raising children would be difficult: a cloned person has no parents and no relatives, so society may treat it coldly—clones would be stigmatised.

Some textbooks argue that, when grown up, a clone could unknowingly create offspring with people who are genetically very close (incest). When the technology of cryopreservation is involved, it would be hard to determine parenthood after many decades (6; 10; 17). This would involve the risk of consanguineous mating, which could pose a threat to the quality of the Chinese gene pool. Some textbooks believe that the progress of humanity partly derives from people's sense of duty towards the household and the family. And without the honouring and nurturing of parents, or filial piety, the next generation would weaken substantially (10; 17). Most textbooks mention the family as the cell of society. According to textbook 10, the environment of a clone is just as important as its genetic make-up. And, if cloning leads to stigmatisation, then reproductive cloning forms a biological time bomb. Textbook 10 argues that even in a communist society, where education is largely taken over by the state, the family household is still the basic unit of society and the state can never take over the task of raising a family.

Other counter-arguments relate to human health risk (4; 6; 10; 12; 15–17) and the ultimate impossibility of copying a complete individual (8; 9; 11; 12; 14; 16). The human risk factor in reproductive cloning here refers to the high chance of technological failure, the short lifespan and the possibly poor quality of life of the clone due to technical problems (10). But it also refers to a failure to respect life and family relations: it violates humanitarianism (rendaozhuyi). Furthermore, as a clone is only biologically a copy of its model, it does not have the psychology, or the behavioural and social characteristics of the model. For this reason, the clone is not considered to be a whole person. It is a person that has lost its 'self' (sangshi ziwo de ren) (15: 90; also 6). And textbook 11 argues that, as a clone is always made for a certain purpose and with certain traits, it will always have a warped mind and suffer from a pessimistic psychology and fate—it may feel that it is a tool in the hands of society (11).

In short, cultural and political concerns about the quality of the Chinese population and the continuation of current Chinese notion of the family are central to the prohibition of reproductive cloning in China.

The bioethical problematisation of human reproductive cloning, then, involves ideas related to the family, notions of what constitutes a healthy Chinese population stock and the appropriate political measures to take to ensure the sustainable vitality of the population. Even though in the 1970s and 1980s political intervention in the life sciences has been foresworn and emphasised in adages such as 'there are no forbidden zones in science', in the new millennium there are socio-cultural concerns, partly specific to China, that require a 'forbidden zone' in science to assure optimal conditions for the development of the population. Ironically, the rationale for this science policy is aimed at protecting the quality of the population, which in itself has been designed to conform to the scientific principles of cybernetics, systems theory and population science (Greenhalgh, 2008). The question arises, therefore, of whether China has now entered the stage referred to as 'reflexive modernity' (Beck, 1992), in which awareness of the problems accompanying industrialisation leads to uncertainties about science as the source of solutions to them.

Although international pressure was exerted on the PRC to legislate the prohibition of human cloning, China did not budge, as it had its own political priorities. Instead of creating legislation, it went ahead and produced authoritative guidelines prohibiting reproductive cloning, with an eye on the possibility of embryo cloning aimed at pharmacogenomic applications. Over time, however, the distinction between reproductive and therapeutic cloning was adopted in the regulation of all countries interested in advancing therapeutic cloning. The presumption that Asian countries would adopt international regulation lock, stock and barrel is therefore incorrect. But so is the contention that Asian countries are, per definition, against adopting bioethical ideas initiated in what is regarded as 'the West'. The official PRC stance highly values therapeutic cloning. This was clearly expressed in China's opposition to the 2005 UN motion against human cloning. As the motion did not distinguish between therapeutic and reproductive cloning, the PRC would not underwrite it. Although China did not legislate against reproductive cloning, the idea was certainly condoned, albeit for different reasons. But it also became clear that the textbooks for medical and bioethics were not representative of Chinese culture in the sense of reflecting an agreement about the notion of reproductive cloning. Although diverging notions of the family, the population and reproductive cloning co-exist, the views that have discussed in this case study are mostly those of the intellectual elite, which generally harbour faith in scientific progress and are closely affined with the official stance of the socialist government.

Case 11: Formal Discussions of Bioethical Standards in Therapeutic Cloning and hESR in China and Japan[3]

hESR is controversial in many societies, and extensive debates have been held on the use of embryos and oocytes in scientific research (Holland et al., 2001; Bonnicksen, 2002). The basis upon which regulation for hESR and therapeutic cloning is created is conditioned by both international and domestic factors, and can be conceptualised in terms of bioethical problematisation. Understanding the bioethical problematisation in China and Japan provides us with insight into the way bioethics discourses connect international, domestic and local realms in various ways. The bioethical problematisations in therapeutic cloning and hESR in China and Japan draw upon the norms and values deployed in official discourse, strategies of decision-making inherent to the traditions of policy-making, and the different priorities of existing science and population policies.

This case study draws on two different studies. The first builds on the study of textbooks for medical ethics as described in Case 10, while the second draws on a study on the national discussion of hESR in Japan. The national discussion (kokumineki giron) on hESR in Japan was said to be barely alive and, as no religious or cultural canons forbid hESR in Japan, it was assumed that national discussion on the normative status of the embryo would be uncontroversial. Paradoxically, according to most of the 50 people I interviewed from April to June 2006—including stem cell scientists, regulators, monks and housewives—debate on hESR is considered to be very important to science policy-making. This case study shows how such attitudes are related to official discourses and how they contrast with those in China.

China

China has a rich tradition of discussions on medical ethics (Unschuld, 1992; Nie, 2005), which it gradually revived in the 1980s, in conjunction with medical ethics ideas imported from 'the West' (Qiu, 1987). Discussions on medical ethics have been mainly confined to exchanges among intellectuals and biomedical professionals, and only very gradually have they started to involve 'the people' in general. In this process, a shift took place from the use of Marxist categories of class to a view of biomedical ethics that recognises the importance of traditional notions of the body and medicine that have reappeared since China's opening up and reforms from the late 1970s onwards. Although human reproductive cloning has been discussed in the press, and has been a sensational topic for novels and magazines, therapeutic cloning has not caught the imagination of the media and has been mainly confined to discussions among academics. The considerations important in the development of the creation of formal bioethical standards I therefore describe on the basis of the textbooks for medical and bioethics. At stake in this discussion is the bioethical problematisation of therapeutic cloning, which, in the textbooks, has been extremely cautious. I discuss the reasons for this at the end of this case study.

Discussions on therapeutic cloning in China and Japan, compared with those on human reproductive cloning, are predominantly low-key. In both countries, as elsewhere, human cloning was rejected by the majority; therapeutic cloning was defended, but with varying intensity and for diverging reasons. When the importance of acknowledging and defining a moral and legal difference between reproductive and therapeutic cloning to the development of genomic medicine became clear, Chinese bioethicists, regulators and scientists began to categorise the issue of therapeutic cloning together with that of hESR in a separate category.

The problems acknowledged and discussed in Chinese debates on hESR mainly cluster around two issues: the value of the embryo and embryo donation. In most cases, however, the views critical in using embryos in research are discussed as if they are problems that only concern foreign countries. Most textbooks did not mention bioethical problems around hESR in the PRC, with the exception of textbooks 9, 11 and 14, which called for caution, as described below.[4] Textbook 10, as an example of a textbook unaware of problems linked to embryo research, states that, in view of the great social and economic advantages promised by hESR, the Chinese Ministry of Health has made it clear that it supports hESR for clinical purposes. In support of this policy, the Southern Human Genome Organization's ethics committee declared that 'medicine is to serve benevolent [use of] technology' (Yiweirenshu).

Although medical textbooks (9, 11 and 14) relate ethical issues of hESR to Chinese society, this did not lead to any ethical stance apart from calls for vigilance and reflection about more effective regulation. Textbook 9 asked whether the treatment of the embryo in hESR could be regarded as benevolent (ren), weighing the violation of value attached to the embryo if regarded as a moral individual (daode geren) against the unethicality of not helping the severely disabled. Textbook 11 was more concerned with safety, while questioning whether one can prevent therapeutic cloning technologies from being used for human reproductive cloning, and whether safety could be guaranteed, even in the case of therapeutic cloning (11: 190). Only textbook 14 related the discussion on the value of human life to a Chinese context, including that of Confucianism, which it describes as an a-religious form of regarding human life as sacred (Tiandi zhi xing, ren wei gui). It is much concerned with its regressive influence on attitudes towards the embryo in society. For instance, it states that up to the present day, many consider it wrong to use the organs of the dead for donation, for families hope that the deceased will be able to enter Heaven (Tianguo) unscathed (14: 116). But textbook 14 argues that, as embryos and fetuses are not social human beings but biological human beings, the fetuses that have been aborted as a requirement of the family planning policy or that are spare embryos after in vitro fertilisation (IVF) treatment can be used in research.

Although some textbooks discuss arguments against hESR, none of them engaged with the issues of the potential emotional or financial pressures on women to donate embryos or oocytes, benefit sharing, the psychological problems related to abortion, attachment to pre-birth life or embryo trade in a Chinese context: they were mainly defined as problems belonging to foreign, Western, feminist or Christian worlds (10; 12; 14–17). So why do the textbooks not hold a discussion about hESR as they did about human reproductive cloning? There are various reasons for this. First, there was genuine unawareness among regulators and medical professionals about the possible problems with embryo donation and the moral status of the embryo (interview Prof. Liang, 28 April 2007) owing to the lack of public debate and discussion at the time. Second, the medical ethics of hESR in the examined textbooks follows the official political line on the subject. In the case of hESR, it means that they do not refer to problems related to the moral value of the embryo in China, as it would contradict official policies on birth planning. And, third, discussions of the prevalence of hurt feelings around abortion and the values ascribed to the embryo, fetus and also organs. Thanks to empirical studies by Jing-bao Nie (2005), we know that a large proportion of the Chinese population regards life as beginning at conception, and that many women experience serious mental and physical problems as a result of the family planning policy. As hinted at by textbook 14, worries exist in Chinese society about the donation of any bodily tissues, which have been imbued with cultural values of ancestral gratitude and karmic respect.

It is clear, however, that ideas on morality and ideology around the embryo are alive and pervasive, and textbooks recognise an ideological (or non-scientific) stance as essential for people's wellbeing and willingness to donate. Thus, textbook 15 states that the effect of the application of new technologies will depend on the quality of medical and bioethics (15: Ch. 1). But the textbooks express great diversity of views regarding the moral status of the embryo in China, varying from precious life to something with no moral value at all apart from as the precursor of 'social human beings'. Official policies seem to support the latter fully and the former indirectly. But the view that fertilised eggs created by therapeutic cloning may be research material for only 15 days, as stipulated by international guidelines, seems to be strangely at odds with the idea that the embryo and fetus are discardable at later stages. Although the textbooks accept that fetuses develop a nervous system and, as a result, have sensations, fetuses are in no way thought to deserve the respect owed to human individuals. Nevertheless, scientists are advised to accept 'foreign' guidelines. This is justified as both a way of preventing reproductive cloning and as a way of conforming to the requirements of the international science community.

This position, however, has not become an official discourse in the media—there have not been great efforts invested in persuading the public of the value of this view or to involve them in the process. The result was a very weak and contradictory bioethics discourse: no bioethical problematisation had been developed to speak of, leaving scientists in a regulatory ambiguity. On the one hand, they had relatively a lot of leeway to experiment with new methods and theories; on the other hand, they had to live with a dubious reputation among both the international science community and 'the people' aware of it. It is therefore not surprising that it was Chinese returnee scientists who started to set up ethical review boards for embryo donation in their hospitals.

Japan

Japan, as a modern, wealthy welfare state with a large state bureaucracy, a high standard of healthcare provision, and advanced level of science and technology prides itself on being a democracy and being in communication with its citizens. The regulation of hESR in Japan was preceded by a period of stimulating public discussion and expert discussion, which took seriously the diversity of opinion in Japan. There is no Japanese equivalent of the flurry of textbooks for medical and bioethics that has come about in China since the late 1980s. To obtain an idea of debates and arguments in official discussions concerning hESR, one can also examine the views of the committee members responsible for discussing the bioethical regulation for reproductive cloning and hESR. It will become clear that, even though the factors conditioning hESR are complex, the bioethical problematisation of hESR in Japan is both more developed and more nuanced than in China.

In Japan, human reproductive cloning is considered as morally wrong, so it was prohibited. So strong was the consensus about its potential dangers that, as an exception, the country developed legislation to forbid it. Most guidelines for research and medical practice are so-called 'soft guidelines', and are backed up by the Japan Medical Association, research societies and the various related ministries, such as the Ministry of Health, Welfare and Labour (MoHWL), and the Ministry of Science and Technology (Ida, 2002). Thus, reproductive cloning was prohibited by the Cloning Restriction Law passed in June 2001.

But Japan, just as the PRC, also intended to provide financial support for and stimulate the development of the life sciences, in particular therapeutic cloning. For some European and Asian countries, this became especially relevant after former US president George W. Bush announced the cut in federal funding for hESR in the USA. To enable therapeutic cloning, new regulation was developed and a new concept was created, the so-called 'specified embryo' (toku teihai), to de-link it both from reproductive cloning and from the issue of abortion (Hishiyama, 2003; Kayukawa 2003). In official policies on reproduction and population health, ambiguous areas exist about the moral status of the embryo and abortion. These areas condition the ways in which the moral order around reproduction is established and the embryo is valued. Thus, although abortion in Japan has not been removed from the criminal code, most of the approximately 300,000 abortions that take place annually are permitted in cases of psychological and financial problems, if the pregnancy threatens the health of the mother, and in case of pregnancies resulting from rape or violence (see Norgren, 2001; Sato and Iwasawa, 2006).

In September 2001, the Koizumi government approved the use of embryonic stem cell lines for basic research. With the subsequent issue of 'Guidelines for Derivation and Utilization of Human Embryonic Stem Cells' (2001) by three ministries, Japan became one of few countries that officially permits research on human embryonic stem cells (hESC).[5] Although it allowed the destruction of embryos, the authorised guidelines emphasised respect for stem cells and the embryo itself, and that informed consent from the donors with the possibility of withdrawal was obtained (Ida, 2002).[6] In a next step on the road to biotechnology regulation, on 23 June 2004 Japan's highest-ranking organisation for science and technology policy-making, the Cabinet's Council for Science and Technology Policy (CSTP; *Kuni no Sogo Kagaku Gijutsu Kenkûkai*) opened the door to the cloning of human embryos (Foreign Press Centre, 2004). Shortly thereafter, on 1 July 2004, the Health Ministry's Ethics Committee voted unanimously to allow the use of stem cells from aborted fetuses in clinical research (Cyranoski, 2005). In line with the Cloning Restriction Law passed in June 2001 with the purpose of banning human cloning, the task of the council was to deal with such matters as the handling of cloned embryos. After three years of debate, the Bioethics Expert Committee (BEC; *Seimei Rinri Senmon Chôshakai*) of the Council decided in favour of embryo cloning for the use of regenerative medicine (Foreign Press Centre, 2004).

The BEC of the CSTP defined the embryo officially as the 'germ of life' (*hito no seimei no hôga*), a concept that accommodates views of people from various religious and cultural backgrounds, and arguments for and against embryo research. The majority of scientists regard this definition as typically Japanese, and contrast it with the religious dogma that views induced abortion as evil. According to law professor and BEC member Ida Ryûichi, the view of the embryo as the germ of life means that its destruction equals killing a potential human. But, at the same time, in a non-religious view, it allows the argument that the use of embryos left over after IVF treatment (supernumerary embryos) could actually lead to something positive, such as new therapies for curing serious diseases.

Although the 2001 guidelines allowed the destruction of embryos, they emphasised that it was necessary for scientists to express respect for stem cells and the embryo, as well as to obtain informed consent from the donors with the possibility of withdrawal (Ida, 2002: 1147–8). Each research application needs to be double-checked, first by the 'local' institutional review board (IRB) and then by the Expert Committee at the governmental level. The cell lines in question may only be used for basic research, and embryos can only be obtained from married couples undergoing IVF treatment, after obtaining written consent and without offering compensation. It is clear that the regulation of the moral status of the embryo, and the procurement of oocytes, is a matter of great concern to regulators. Not only can this be surmised from the regulation that actually begins with the prescription of respecting the embryo, but also from the heated discussion within the CSTP's BEC, which had been organised to reflect a diversity of views.

Professor Hamiguchi from the University of Tokyo, and member of the BEC, was one of the main opponents to the final recommendation of the committee to allow embryo cloning in Japan. At a time when scientists pointed out Hwang Woo-Suk's success with embryo cloning in Korea, and the government followed their advice, Professor Hamiguchi had already realised that there had to be something wrong when so many oocytes had been collected for research. He also knew women in the so-called

Anti-Eugenics Network who were angry about the government's plans to go ahead with hESR, even though a proper debate had not yet been held (interview with Prof. Hamiguchi, 30 May 2006). Professor Hamiguchi spoke of the need for a global debate. For, if China and India were to go ahead with hESR, he said, it meant an unfair (fusei) situation in scientific development. Another problematic matter, according to Professor Hamiguchi, is that, even within Japan, the Jôdo Shinshu (Pure Land Shin Buddhism) has a strict notion of abortion.[7] It is based on the importance of life (inochi no taisetsusa). But, in their case, Professor Hamiguchi suspects the ulterior motive of stimulating Japanese birth for reasons related to colonial expansionism. This, Professor Hamiguchi argues, illustrates the wish of the modern nation-state to control reproduction. Agreeing with the feminist movement (Sôshiren) and some religious groups, especially Omoto Shinrikai, Professor Hamiguchi opposes state control over reproduction (interview, 30 May 2006). As a BEC member, Professor Hamiguchi has attracted much attention as an opponent of hESR, and as a troublemaker for science. Various people that sat with him on committees became annoyed by his 'unscientific attitude' (interview with Dr Fukuda, 31 May 2006).

In July 2004 the BEC decided to allow embryonic cloning after the creation of adequate regulation for oocyte donation and the construction of a solid scientific basis. But the cloning scandal surrounding Hwang Woo-Suk in Korea, in December 2005, and the lack of progress made internationally, meant a setback to the advancement of the regulation. In May 2006, the CSTP decided that although embryo cloning would be allowed, it could only actually take place after sufficient experimentation on primates. And although some scientists complained that this was tantamount to prohibiting human embryo cloning, as the cloning of monkey embryos is regarded as very difficult, others argued that the current regulation has prepared Japan for sudden international advances in human embryo cloning. As a preparation, regulation for oocyte donation was created. Couples were to be allowed to volunteer spare eggs, left over after IVF treatment, for the purpose of hESR. To this purpose, the CSTP decided to create the job of co-ordinator, the role of whom is to negotiate between the researcher/hospital and volunteers. According to Professor Hamiguchi, these moves were entirely in agreement with the two conditions formulated in July 2004.

In Japan, the political situation around the status of the embryo is extremely complex. Although respect for unborn life and the dignity of the embryo are officially prescribed in case of research embryos, aborted embryos up to the age of 12 weeks may be treated as rubbish, and are used in fetal research. A central source of confusion regarding hESR lies in its official encouragement. It is a contradictory situation in which hESR aims to select the best embryos for research, even though abortion is criminalised in Japan, and even though embryo selection through abortion and pre-implantation genetic diagnosis (PGD) is prohibited in most cases.

Although complex, discussion on the status of the embryo in relation to hESR in Japan has been held by an expert committee constituting members of various diverse scientific, political and stakeholder backgrounds. The purpose of this strategy was to gain a diversity of views and to generate ideas for creating consensus in the regulation of hESR. The resulting regulation, although taking into account the sensitivities around the status of the embryo in Japan, was not conducive for hESR, as will be discussed in Case 12.

Bioethical Problematisation in China and Japan

Comparing the bioethical problematisation in China and Japan, we find a difference in orientation to, and different structuring of, participating publics. Discussions in the media on the development of regulation in China were not encouraged, while in Japan the public could follow each stage of it in the newspapers and on ministerial websites. Textbooks for medical and bioethics in China show that the government regulation on embryo research and therapeutic cloning was not built on a multiplicity of views in society regarding the moral status of the embryo, while the Japanese government's strategy aimed to organise public debate as an expedient for creating consensus by playing out radically diverging views in society in the BEC. While China's regulation of hESR gave priority to the facilitation of life science research, in Japan emphasis lay on the need for consensus and the inclusion of all relevant views, including those of minorities. This strategy aimed to be inclusive of relevant stakeholders (including women and handicapped groups) and to prevent scandal in the long run. Bioethical problematisations in both countries build on contradictory sensibilities among the population. Thus, in China, the one-child policy has emphasised the 'quality' of children in terms of health and education, and encourages attachment to the embryo and fetus, while at the same time compelling advice to abort fetuses can cruelly break that tie; in Japan, although abortion is a widespread option in dealing with an unwanted pregnancy, it has not been taken out of the criminal code, implying that abortion is only tolerated. And while in Japan, for the purpose of scientific research, the most suitable embryos are selected, the same is not allowed for PGD or IVF.

The bioethical problematisation of hESR in China and Japan set out to facilitate hESR, but was characterised by different priorities, resulting in a weak and strong bioethical problematisation. While the broad rationale behind China's weak bioethical problematisation was to improve both China's international science position and the quality of the population by involving experts in regulatory efforts in an ad hoc fashion, Japan's strong problematisation was the result of a strategy of consensus through open debate to include the interest of minority groups that were considered to have socio-political leverage in an attempt to encourage hESR rooted in broad social consensus.

Case 12: Global Moral Economies and Bioethical Articulations of hESR in China and Japan[8]

Global discussions on science and policy-making have generated awareness of a need for science to be ethical and culturally appropriate, of a need to elicit the co-operation of society. Having become influential, bioethical arrangements have indirectly come to give direction to developments in research, have become part of the institutional make-up of the life sciences and have become of strategic value. Thomas Gieryn's concept of boundary-making (1983, 1999) showed how scientists, through 'boundary-making', try to expand their authority to other knowledge fields, monopolise professional authority and protect their autonomy over professional activities. But with the development of bioethical research regulation, a new situation came about in which international and domestic views on moral life issues have become essential to the shaping of life science practices.

Examples of 'boundary-making' in stem cell research, and 'good research practice' show how hESC scientists in China and Japan articulate what they regard as international bioethics in their own practice through the development of their ideas about the ethics around hESR and therapeutic cloning. Elite scientists are aware of the views of 'Western' governments on hESR and the world view of East Asia as being 'unethical'. Most are also aware of the various stances on the moral status of the embryo in their own country. They forge their own views on the basis of the differences they observe among these international and domestic views. The creation of these views involves delineating their stance on good research practice from those they do not agree with or want to put up as a foil against what they regard as 'good' research.

Such boundary-making may serve various purposes, among which are to advance their interests in the field, to justify resource allocation and to build confidence in their field of expertise. Although it is impossible to say to what extent they succeed, an examination of boundary-making sheds light on how the aims, interests and needs of scientists are packed in various culturally-specific strategic forms. The concept of culture here is used in a broad sense, indicating the ways in which bioethical science practices are given meaning through scientists' strategic performances drawing on norms and values (Baumann, 2002) that vary across geographical localities and socio-political interest groups. A discussion of boundary-making among Chinese and Japanese stem cell scientists will show that different roles ascribed to international, domestic and local research practices calls for a distinction between internal and external boundary-making.

This case study draws on two studies. The first builds on a study of stem cell scientists and boundary-making in China. It uses more than 60 semi-structured interviews, of which 44 were with scientists involved in stem cell research and others with policy-makers, medical doctors, nurses, regulators, monks and female workers. Interviews took place over a period of ten weeks from March to May 2007, and over another nine weeks from October to December 2007. Most of the scientists interviewed belong to well-known science institutions, Chinese Academy of Sciences, Chinese Academy of Medical Sciences, Peking Union Medical College, Beijing University, Fudan University, Zhongshan University (Shanghai), Shanghai Jiaotong University, Zhongshan University (Changsha), Renmin University (Guangdong), Hainan Reproductive Medicine Centre and Huazhong Keji University, Wuhan and various hospitals, such as Ruijin Hospital in Shanghai, and Zhongshan first and second affiliated hospitals. The stem cell scientists and physicians interviewed are affiliated with institutions of varying status and reputation. Interviews were conducted in both Chinese and English, with the language used being decided by the interviewee (translations are indicated in the text). The second study is the one referred to in Case 11. I have used pseudonyms for the interviewees in both cases.

China

Asking Chinese hESC scientists about ethical science yields various references to practices associated with 'good' scientific behaviour. These references place Chinese science solidly in international, domestic and rural research practices. But much more affiliation was expressed with the ethicality of international research practices than with those observed within China. Interviews with Chinese scientists who expressed their views of what ethical hESR is clearly indicate this.

Chinese stem cell scientists generally share notions of 'bad science' that refer to China's history of pseudoscience during and after the Cultural Revolution (1965–1975), including Lyenkoism Traditional Chinese Medicine (TCM) (Sleeboom-Faulkner, 2008a). All 44 stem cell scientists but one relegated any pluripotent stem cell research based on TCM to the status of pseudoscience, and were keen to make clear that evidence-based research is central to *their* science, expressing allegiance with 'universal' science. All stem cell scientists I spoke with condemned the clinical application of untested stem cell therapies for large sums of money, such as the stem cell practice of Huang Hongyun in Beijing, who applies olfactory sheathing cells from fetuses to patients. Selling experimental therapy that cannot be expected to cure patients is also regarded as unethical. Also, scientists who said that they, in principle, would allow experimentation in desperate cases rejected the exorbitant prices therapists ask for. Thus, Professor Duan from Guangzhou opposed the use of mesenchymal or derived neural stem cells to help people with spinal cord problems in Shanghai and Beijing simply because patients pay bills for a bad method. Basically, then, what was regarded as unethical was asking for money in return for experimental stem cell research, regardless of whether this was done with fetuses, oocytes, embryos or through unscientific methods. Knowingly selling something useless, rather than selling unethical science, was rejected here.

The notion of unethical science, rather than science associated directly with embryos or fetuses, was associated with scientific backwardness, contrasted with elite science in sophisticated laboratories in metropolitan and cosmopolitan science hubs. Those with research experience in scientifically advanced countries were regarded as the most advanced and ethical, followed by scientists in elite laboratories; below were situated the scientists ensconced in rural laboratories in the provinces. Scientists with international ambitions conflate 'bioethical' and 'good' science, which they contrast with 'unethical/bad' science threatening the research field. Among elite scientists it is a concept of science without due bioethics review that signifies science that lacks a good research method.

Most scientists presumed that researchers that have experience in an internationally well-known laboratory acquire proper scientific training, and learn about the norms and standards that make science successful. Success, here, is equated with having publications in international, peer-reviewed, high-quality science journals. Researchers that have enjoyed education abroad use this boundary between themselves and those educated in China to indicate their ethical awareness and capacity to work in an ethically sound way. But domestically-trained scientists claim that scientists that have been abroad are not necessarily good scientists, and make clear that having international publications does not guarantee the bioethicality of a scientist. Professor Yang, in Wuhan, lamented that experience abroad makes researchers bioethically savvy so that they know how to dodge the guidelines in more cunning ways. He sighed that scientists with their names on international publications leads to absenteeism, and makes them drive their laboratories too harshly. He partly blamed the government for not controlling unethical practice in hESR and for wasting money (interview 26 November 2007). Other scientists blamed foreign returnees for wasting the large sums of government funding they were lured back to China with. Others still point the finger at the hundreds of fake researcher-cum-therapists who forge deals between research centres and hospitals to gain research funding and permission (interviews Li 30 April 2007; Zhao 17 May 2007). An opponent of hESR in Beijing blamed the government for closing its eyes to unethical practices: 'you can arrest one million political dissidents in one night, but you cannot even control this...!' (interview Professor Peng, 29 April 2007).

Stem cell researchers clearly use government vetting and their affiliation with large university hospitals in the metropolis to justify their own research as ethically correct. Thus, embryologist Professor Li, from Guangzhou, says: 'Recruitment for oocyte collection is not allowed. Perhaps smaller clandestine centres engage in this kind of work in rural areas. If the Ministry of Health closes them, they may not care about it. But the larger hospitals have their own resources, and cannot afford to do this'. The practice of bona fide science in large metropolitan hospitals and universities is routinely contrasted with the practices of small clinics in the country or clandestine research centres. These centres are thought to engage in clandestine embryo and oocyte trade, unlike the large hospitals and stem cell research centres in the metropolises.

In short, Chinese stem cell scientists routinely make a distinction between foreign ethical, domestic and rural unethical practices. Whether these practices are based on bad scientific method or on the violation of bioethical guidelines scientists do not usually point out. In their ethical boundary-making, stem cell scientists identify with foreign research practices and condemn rural ones, while invoking government approval, funding and licensing to confirm their research is up to scratch. In China, then, a distinction between internal and external boundaries is significant to the identity formation of scientific institutions and bioethics.

Japan

In Japan the notions of experimental research and unproven method are more clearly delineated from the ethicality of research. Here, the categorisation into international, domestic and rural research plays a less important role. References are made to the USA and the UK, whose bioethical regulatory arrangements, it is pointed out, have not prevented hESC communities from flourishing. References are also made to China and South Korea, whose tarnished reputations as rogue science countries have not stopped them from gaining an increasingly higher reputation as globally advanced scientific powers. But when it comes to boundary-making, discourses among scientists on the ethicality of stem cell research predominantly involve clashes between interest and pressure groups at home.

Despite the fact that Japan has created regulation to allow the isolation of embryonic stem cell lines and therapeutic cloning, leading scientists such as Nishikawa Shinichi, Akabayashi Akira and Okano Hideyuki have voiced their dissatisfaction with the strict regulatory regime in Japan. Although guidelines allowed the destruction of embryos, the need to respect hESC and the embryo itself was emphasised. Next to obtaining informed consent from the donors with the possibility of withdrawal (Ida, 2002: 1147–8), each research application was to be double-checked, first by the local IRB, and then by the Experts Committee at the governmental level. The cell lines in question could only be used for basic research, and fertilised eggs could only be obtained from married couples after written consent and without compensation (Harris, 2002). Although in the UK scientists may regard 'clear guidelines and strict rules' for hESR as work-enabling (Wainwright et al., 2006: 742), in Japan a combination of 'unclear and strict guidelines' is perceived to have a restrictive effect. And, in contrast with the situation in the UK, in Japan knitting together bioethical activity with research needs has elicited strong protest from scientists in the form of letters of complaints, publications and pressure on governmental institutions for the clarification of bioethical regulation (Hishiyama, 2003; Kayukawa, 2003; Slingby et al., 2004).

The strictness of the regulation of hESR in Japan was further illustrated by the fact that the application for permission to use stem cell lines was extremely laborious, and every change in the project and personnel required additional applications. Both senior and junior Japanese embryonic stem cell scientists expressed their envy of UK research assistants, who can handle stem cell lines without the need to apply for extra permission. As a result, they argued, Japan's hESR is lagging behind—Japan has a high success rate in mouse research and a very low success rate in hESR, as measured by Japan's global share in publications (Cyranosky, 2005). But although less strict regulation was demanded, none of the scientists interviewed expressed a wish for a split from federal- and state-based regulation, such as can be found in the USA, or a slack regulatory implementation, which is associated with China. Scientists seem to be in solid agreement that reputation and trust are regarded as far more important to their institutional survival than 'over-permissive' regulation (Sleeboom-Faulkner, 2008b).

One of the problems indicated by scientists is the mandatory respect for embryos and embryonic cells in the laboratory. The governmental subcommittee on hESR defined the human embryo as the 'sprout or germ of life' (Seimei no hyôga), demanding respect in handling human embryos and hESC. The disagreement in Japan on the bioethical regulation for hESR applied in a clinical setting from 2003 to 2006 provides us with a discursive background on existing views on the fetus. Nakayama, a professor and principal investigator in developmental paediatrics at Kyoto University, related how the regulatory committee (under the MoHWL) did not succeed in formulating the guidelines on the regulation of hESR in a clinical setting, partly owing to public consternation over two fetuses discovered in the waste of Yokohama City Hospital (Yokohama Shinai Byôin) in the summer of 2004. According to Nakayama, people are afraid of scandal and change: Japanese people like to 'celebrate the emperor' (Goshô ni iwai); they prefer to obey their superiors or the government, rather than trying out something new.

Professor Kato, a principal investigator at Keio University, believes that the delay in the guidelines was due to feminist anti-abortion groups, who, he claims, argue that scientists drive up abortion rates. This accusation has historical roots. In 1948 the Law for the Protection of Mother and Child allowed induced abortion. It was even listed as one of the three forms of birth control (sanji seigen) (Norgren, 2001: 83). It was only in 1952, against the backdrop of the baby boom, that contraception was encouraged, but with little effect. Even today, the main means of contraception in Japan is the condom, while the pill, introduced in Japan in 1992, is used very little (Sato and Iwasawa, 2006; Tsuge 2010). Predictably, Japan's abortion rate is high: in 2000, a quarter of all married Japanese women indicated that they had experienced induced abortion (Sato and Iwasawa, 2006: 38). Now, Professor Sato hopes that regulation will protect the reputation of scientists against accusations sometimes heard of their thirst for fresh embryos.

Professor Suda, originally an embryologist, but now a professor in bio-ethics and a member of the Committee for the Creation of Guidelines for the Use of Human (Somatic) Stem Cell in a Clinical Setting, opposed the views of Professor Kato. Professor Suda urges women in Japan to ask whether it is truly for science or to please the physician and others around them that they want to donate their embryos and oocytes. Professor Suda also believes that, as stem cells are little understood as yet, it is far too early for human experimentation. As an embryologist, Professor Suda looks back on how much fun it was to conduct experiments with sea urchins:

> It was really interesting. Trying out new things was very exciting! It is a kind of curiosity that motivates scientists. I believe that there is a lot of this curiosity and challenge of doing things that no one has ever done before, even though they may also do it for the patients (interview Suda; 27 May 2006; transl. by Sleeboom-Faulkner).

Someone needs to limit this boundless curiosity, Professor Suda argues, as scientists are not going to stop themselves. But Professor Suda does not believe that the government is going to, as she does not think that the Ministry of Education, Culture, Sport, Science and Technology would risk its position in global competition—its priority is getting Japan out of the economic slump.

Professor Shimada, head of the Translational Research Centre at a university hospital, has similar, and perhaps even more outspoken, views about the safety dimension of embryonic stem cell research. Embryonic and fetal stem cells, according to Professor Shimada, are 'still too little understood to inject into the brains of patients.' Professor Shimada blames the laxity of government regulation for the poverty of the discussion and the unclear status of the embryo/fetus. Like Professor Suda, Professor Shimada asks for stricter rules 'to restrain ourselves'. This, Professor Shimada expects, will improve the quality of bioethics committees, which currently varies greatly between universities.

In Japan, scientists arguing about the regulation of hESR were much more concerned with internal than external boundary-making, and a much clearer line was drawn between bad science (unevidenced) and bad ethics. Although scientists in Japan are aware that 'bad science' takes place, no one believes that the government would knowingly tolerate it. The main issues, therefore, are those of the status of the embryo, embryo and oocyte donation, and the strictness of the implementation of bioethical guidelines. Japanese scientists are expected to go to meetings on bioethics, be aware of the regulation and follow strict procedures when applying for research funding; these, as is often remarked, are stricter rules than those imposed on scientists abroad.[9]

In spite of the diverse perspectives on hESR, principal investigators all agreed on the need for regulation of the bioethical aspects of research, both to protect the scientists' freedom of research and to protect scientists against their own curiosity and creativity. Japanese scientists, then, have to work on themselves to do their research ethically. The reputation of scientists is inherently linked to this ethicality, and is crucial for them to receive research funding and, in many cases, employment. All scientists interviewed engaged in discussion with domestic standards of bioethics regarding hESR, even though they might prefer foreign, looser ones.

In short, Chinese and Japanese discourses among scientists around the ethics of hESR are shaped differently. Chinese discourses do not make a clear distinction between bad science and bad bioethics, and mainly use the West as a comparative reference in their bioethical boundary-making, while 'rural practices' are used as a foil. In Japanese discourses the distinction between bad science and bad bioethics is clearly linked to different discourses, and all hESC scientists are well versed in bioethics discourses. Although bioethics discourses in Japan do refer to the pressure of what is regarded as Western bioethics, the contestation of bioethical hESR took place mainly domestically through arguments formulated by feminists, disabled groups and religious sects, and taking into account the views of humanists and scientists, allowing bioethics to become rooted into Japanese society.

Conclusion

This chapter set out to understand the three political stances towards bioethics: that globalisation leads to a universal, homogeneous bioethics; that bioethics is driven by oppressive, neoliberal forces; and that bioethics should have, and has, Asian roots. The discussion of three cases shows that such division between neoliberal, anti-neoliberal and nativist perspectives does not facilitate a better insight into the development of bioethical practices. The case study of bioethical attitudes towards human cloning in biomedical textbooks in the PRC (Case 10) clearly showed that critical discourses borrow from international ethics discussions, which are interpreted and translated into a domestic context, where a limited number of positions is defended for reasons entangled with the domestic policies and the socio-cultural and economic circumstances belonging to the relevant life assemblage.

More helpful is the notion of bioethical problematisation, as it links official discourses to relevant policies and institutional histories, and provides an indication of to what extent sections of the population are engaged with bioethics at home and bioethics regulation. Thus, in the case of China's hESR, it was more relevant that bioethics fitted in with science and population policies than with rooting the bioethics in 'indigenous' views. It became clear that a limited number of experts are influential in defining what is ethical in the context of Chinese institutional development—they translate what are regarded as international problems into bioethical problematisations valid to Chinese regimes of governance. In Japan, by contrast, bioethics has a longer history and internationally relevant problems of bioethics have become more embedded in the current practice of hESR. It was in a domestic embedding that the development of hESR was constricted as a result of stringent guidelines, and contradictive notions of the moral status of the embryo exist side by side. The notion of bioethical boundary-making indicated how Chinese scientists compare their practices with what they regard as Western bioethics in contrast with local 'backward' bioethical practices, orienting themselves outwardly, while Japanese scientists look inwardly to boundaries between domestic stakeholders, including civil society groups. These different forms of problematisation and boundary-making are illustrative of the diverging life assemblages relevant to human reproductive cloning and hESR in China and Japan.

6
Scientists and the Public in East Asian Life Assemblages: Risk, Debate and the Professionalisation of Bioethics

Human embryonic stem cell research (hESR) has been controversial in many societies owing partly to the existence of diverging views on the normative value of the embryo, and on the ethical derivation and use of embryos and oocytes required for research. This kind of controversy has been, and still is, expressed in the form of protest movements and organisations that aim to deal with the perceived risks of hESR. In the UK, the Warnock Report, which reacted to the particular problems associated with embryo research, recommended the establishment of the license authority, which materialised in the shape of the Human Fertilisation and Embryology Authority (HFEA) in 1990. This included an attempt to measure the public barometer to attain publicly acceptable regulation (Franklin and Roberts, 2006: 3–5). The HFEA devised bioethical guidelines that would regulate the sourcing of biomaterials and the behaviour of stem cell scientists, while taking into consideration shifts in views among the public. Such bioethical regulation is also important to funding agencies, such as the Wellcome Trust and the European Framework Programmes, who wish to protect their reputation and do not wish to subsidise research that is regarded as bioethically irresponsible.

A standard narrative about bioethics in some Asian countries is as follows. In the Asian societies that are thought not to have traditions of public discussion, governments would not engage the public in deliberation. In these societies, autocratic modes of decision-making would be in charge of regulating bioethics, making public discussions on bioethical issues obsolete. This then would have led to amoral economies taking advantage of the 'bioethical vacuum' that opened up after former president of the USA George W. Bush announced a ban on the federal funding of hESR. Asian governments, then, would use propaganda and rosy visions of promissory science to elicit co-operation from their popu-lations and to stimulate the life sciences, leaving bioethics discussion just to 'establishment intellectuals' rather than to the public.

Both views about the development and functioning of bioethics in the UK and in Asian countries require critical attention. Critics have pointed out that existing traditions of public consultation in the USA and Europe do not always work in agreement with what is advocated. In her article on the HFEA's public consultation process on hybrids and chimeras, ethicist Françoise Baylis argued that the HFEA's public consul-tation process regarding the ethical and social implications of creating human–animal embryos in research was flawed (2009: 43). First, the HFEA failed to provide the public with an impartial assessment of the policy options. Second, the HFEA did not undertake the public consulta-tion so that the public would inform the final policy choice. And, third, the public consultation was not conducted in the mode of genuine com-municative interaction action. Instead, the consultation was rule-guided and strategic.

Critics usually reject modes of decision-making that do not engage the public in the bioethical regulation of the life sciences as autocratic. But in some countries the presumption that the normative value of the embryo should be discussed is being questioned in the first place. Thus, central issues in the debate on the ethical deliberation of hESR are whether coun-tries should adopt internationally acknowledged standards, whether all citizens of a country should be encouraged to engage in discussions on bioethics and whether socio-political movements exist that counter the political hegemony of a universalist bioethics. This chapter discusses modes of representation in public and political discussions on bioethics, how they affect policy-making on hESR in China, Japan and South Korea, and the ways in which scientists in these countries deal with the idea of public influence on the regulation of their research domain.

The chapter analyses the views of scientists and policy-makers on pub-lic debates of hESR in terms of risk. The analytical value of the concept of risk gained ground in social science discussions in the 1980s, indicat-ing that 'modern Western countries' had become aware of the inability to calculate the risk of economic growth and industrialisation to the environment (Beck, 1992, 2007; Adam, 2007). For this reason, these countries were placed in the historical stage of 'reflexive modernity'. But although in 'Western' societies political and intellectual discourses take into account developments detrimental to health, welfare and the environment, policy enforcement continues to take place on the bases of profit estimations, risk and insurance strategies (see Cooper, 2010).

In a way, bioethics institutions have also come to function as a buffer against scandal, critique and loss of reputation among potentially harmful and immoral life science enterprises.

Examining notions of risk in the bioethical thinking of stem cell scientists in mainland China brings us closer to an understanding of the relationship between scientists, global science, the public and the state in the life assemblage it is part of. Using the concept of national risk signature, Case 13 discusses the way scientists interpret bioethical regulation and public discussion in various terms of risk to career, institute and country. The considerations of these scientists are not just coloured by their personal values, but also derive from the position they have in society and their awareness of the comparative importance of bioethical regulation in different countries. A life assemblage, then, is not bounded by territorial boundaries, but by both the perceived and real relations, networks, critique, rules and ideas that interlink a certain research object, here hESR, to the material conditions of its practice. The case study shows that while in one country scientists may regard public discussion as a threat to the development of the life sciences, in another it might be a sine qua non.

On the role of civil society groups, scientists in the public deliberation of hESR and the professionalisation of bioethics in South Korea, Case 14 raises the question of whether the institutionalisation of bioethics is a development just of benefit to the public in South Korean notions of civil society. The case study shows how in some cases the institutionalisation of bioethics may lead to exclusion of the public from significant involvement in the creation of bioethical guidelines. The presumption that standardisation in the fast changing field of bioscience can be left to professionals may have to be questioned: professional bioethicists, if not required to mediate public discussion, risk becoming part of a larger bureaucracy and becoming subject to other forces than that of facilitators of democratic exchanges on issues relating to science and society between public and scientists. It is the particular history of the bioethics movement in a South Korean life assemblage that enables us to see the difference between activist and professional bioethics.

Case 15 on hESR and induced pluripotent stem cells (iPS) in Japan shows how public pressure can lead an otherwise cautious government to go for a risky venture. iPS are a type of pluripotent stem cells made from somatic cells by inducing the expression of specific genes. Compared with human embryonic stem cells, somatic stem cells are easy to acquire, are less prone to immune rejection and are not as ethically controversial, as they do not require the destruction of an embryo. Scientists involved in the research, along with the Japanese government, hope that iPS will soon facilitate clinical applications in human diseases. In Japan, risk-evasive, strict regulation of hESR would have driven life scientists to engage in a form of science involving less bioethics controversy. The science that led to the developing of iPS has often been quoted as an example of ethical stem cell research (SCR). But, from a different perspective, scientists have also regarded the sudden leap into iPS as a high-risk strategy, inviting behaviour that can lead to over-optimism, over-investment and too much pressure on Japanese scientists.

A distinction between notions of the risk, danger and costs of new biotechnologies is shown to be conducive in understanding science innovation in the context of life assemblages in East Asia. Insight into different aspects of the roles of scientists and the public in these three cases show that the relation between policy-making and public discussion is important to the direction in which life science innovation develops. It becomes clear that the absence, presence and the kind of public discussion taking place is closely related to the ways in which pluripotency is taken forward.

Case 13: National Risk Signatures and Public Discussion in China[1]

Bioethics discussions of international science collaborations and debates on the 'risk' of applying new technologies often treat bioethical regulatory issues as a matter of morality (Döring and Chen, 2002; Sleeboom-Faulkner, 2004, 2008a; Qiu, 2006;). But social science approaches indicate the influence of historical factors and political economy on attitudes towards risk. Thus, Ulrich Beck's (1992, 2007) notion of reflexive modernity states that the central problem of 'Western' societies is no longer the 'distribution of goods' (wealth, employment in conditions of scarcity), but the minimisation of risk as exemplified in the admonitions of environmental movements. Despite these indications that risk interpretations of new technologies vary between societies, the idea that basic bioethical guidelines for SCR should be followed internationally is widespread. For instance, the guidelines of the International Society for Stem Cell Research (Daley et al., 2007) presume that moral and bioethical considerations are shared equally on a global level. To put into perspective the difficulties related to such recommendations in the context of a large low- and middle-income country such as China, this case emphasises some of the material aspects of the risk perception of hESR, examining the readings by Chinese stem cell scientists of the risks associated with SCR and the creation of bioethics institutions.

Eschewing approaches that understand bioethical risk as a mere matter of morality or as a social construct, this case study emphasises the materiality and strategic reasoning of bioethical views on risks associated with hESR. Such an approach allows the identification of forms of risk rooted in the everyday practice of Chinese hESR, including moral risk (as a violation of cultural values), material risk (in relation to the distribution of material resources and wealth), political risk (in terms of the political economy of bioethics and public debate) and reputational risk (in terms of personal and national honour). My analysis builds on Tom Horlick-Jones's concept of risk signatures of new technologies (2007), which emphasises the capacities of different technologies to engender and delimit the particular social and cultural interpretations of the risks they generate. But this case study reveals the existence of a certain global awareness among stem cell scientists of risk signatures (Sleeboom-Faulkner, 2010b).

Without taking into account stem cell scientists as reflexive agents, an observer would miss the crucial point that scientists act upon their knowledge of a number of risks identified in practice, that is, the non-sociologists' recognition of risk signatures. In fact, this study on scientists' views on and experiences with hESR in China found great awareness of the institutional constraints, the availability of human biomaterials, laboratory equipment, research funding and the sensitivities regarding hESR. Ample research on hESR in Europe, the USA and elsewhere in Asia yielded similar findings (Bender et al., 2005). This case emphasises that among scientists a global awareness and reflection exists of the different national risk signatures of the 'same' technology, that is, hESR, in different global institutional settings. This means that the strategic reasoning of stem cell scientists at work is informed by the global conditions in which various risk signatures develop; an awareness of the interplay of a manifold of local risk constructs engendered and shaped through the material realities of risk signatures.

In a low- and middle-income country such as China, however, the risks involved in hESR differ from those prevalent in affluent societies. Owing to the varying local circumstances for conducting hESR, including varying permissiveness of regulation and supervision, SCR in China takes on different forms. Compared to affluent welfare societies, where it is in the interest of the researcher to more or less follow national regulation and institutional conventions, in China considerations of reputation, being state-of-the-art and funding eligibility are important in deciding whether to establish costly bioethics institutions when planning a hESR project.[2] In other words, the risk signatures associated with hESR in China and in affluent welfare societies differ owing to, first, varying risk perceptions (social risk constructions) and awareness of hESR; second, a 'real' tangible difference in institutional and material provisions and safety standards; and, third, the different socio-economic circumstances in which oocytes and embryos for hESR are sourced. Stem cell scientists are highly aware of such differences. In China, relatively many scarce resources have been invested into the life sciences, most of the resultant products of which are not expected to benefit the health of the Chinese population. This policy is also controversial among scientists (Blumenthal and Hsiao, 2005; Sleeboom-Faulkner and Patra, 2008). When couples and individuals with little or no healthcare coverage are asked to contribute to the development of the life sciences by 'donating' oocytes and embryos, or by participating in research as experimental subjects, Chinese patients are confronted with different sets of considerations and interests compared with patients in affluent welfare societies. The risk signature of hESR, then, is partly shaped through such sourcing.

An important factor giving shape to the way hESR in China is organised is that public debate on hESR is hardly recognised as a part of the policy-making on hESR. In the view of most scientists, public debate on SCR in itself is considered as a political risk that could potentially undermine hESR in China. This case shows how public debate is part of China's national risk signature for hESR. This study draws on two years of research on international science collaborations in SCR in China, from 2006 to 2008, which involved fieldwork, archival research and interviews with more than 40 scientists from various stem cell laboratories, hospitals and research institutes in China. As this chapter aims to discuss the dynamics of risk perception in bioethical regulation, and not persons, the interviewees in this case have been made anonymous. Although some of the interviewees indicated that they could be quoted in public documents, it is not the aim of this article to attract attention to the interviewees as individual researchers. Approximately half of the interviews with Chinese scientists were conducted in standard Chinese (putonghua); the other half were conducted in English. I have used pseudonyms for all interviewees quoted in this case study.

Discussing public debate with stem cell scientists in China requires clarifying the general perception of the term among intellectuals. Asking stem cell scientists about public debate (*gongkai de bianlun*) on hESR, it becomes clear that the concept of 'public debate' is not necessarily associated with the broad public. Responding to the question 'does China have public debate on SCR', the following answers were common:

Yes, I think so. I went to a conference where there was much discussion on this (Zhai; transl. by Sleeboom-Faulkner).

You can look on the web. There you can see…It is interesting for the college students. Here, people live together in the dormitory. They can discuss this overnight (Bing; transl. by Sleeboom-Faulkner).

Indeed, the majority of scientists pointed out that public debate could be found on the Internet and in newspapers, mainly pertaining to debate among intellectuals. There is a sharp division in the eyes of scientists between scientists and the man on the street (*laobaixing*). Asked whether ordinary people should have a say in creating bioethical regulation, there is great variety of opinion among scientists. Some views are encouraging of public debate:

Ordinary people must also have the right to talk about it. They have the responsibility to talk about this (Bing; transl. by Sleeboom-Faulkner).

Some views indicate strong doubt about the wisdom of involving the masses:

I think they also need a say, but it could lead to forbidding stem cell research like in the USA (Shi; transl. by Sleeboom-Faulkner).

Some views doubt the ability of the masses to understand the main issues:

They might lower the level of discussion (Pu; transl. by Sleeboom-Faulkner).

Some scientists worry that the *laobaixing* will be cheated:

If we explain exactly what stem cells do to all the people, there are some people who use it to make money. They say that their medicine consists of stem cells and make money with it (Wu; transl. by Sleeboom-Faulkner).

Some views doubt the interest of *laobaixing* in the discussion:

They have more important things to do, such as looking after themselves, making money (Su, transl. by Sleeboom-Faulkner).

They are more interested in the price of houses (Lai; transl. by Sleeboom-Faulkner).

The vast majority of scientists did not count the views of non-intellectuals as part of public debate. When asked to think about reasons for including them, the vast majority of stem cell scientists found reasons to exclude them, while it occurred to none that public trust in science could be important. Stem cell scientists in China, rather than the social 'alter-ego' of a critical public, focus on the government as a source of criticism and appreciation. This has consequences for the way they evaluate the role of the public in bioethical issues of hESR, and to what extent they regard the public debate as a risk factor. The national risk signature of hESR in China then proceeds from a situation in which public debate is kept at bay. As scientists are aware that in other countries controversies have resulted in tighter research regulation, the majority of stem cell scientists argue at great length why the public should have no voice in the debate.

It should be emphasised that these opinions represent the views of interviewed stem cell scientists and are not necessarily those of the *laobaixing*. For the *laobaixing* interviewed on this topic (mostly manual workers in the city), although in possession of only a little knowledge about SCR, many were very interested in the ethics of embryo and oocyte donation, and the creation of human clones and cybrids.[3] The question arises here of whether scientists echo government views in underestimating the ability and interest of the masses in matters of science regulation. When asked if there should be public debate about hESR in China, scientists expressed a conditional need for public debate on SCR:

If it would be conducive to creating clear guidelines: So that we know what we must do to do our work. So I think we should talk about these issues (Jing; transl. by Sleeboom-Faulkner).

If the public is educated: But if everyone starts to get involved in it, it may be difficult (Lai; transl. by Sleeboom-Faulkner).

If official authorities channel the discussion: In China you have the People's Congress. They have representatives that can deal with the issues of the people (Sun; transl. by Sleeboom-Faulkner).

If the opinions can be standardised: There are also people who oppose therapeutic cloning. There are various opinions. We try to normalise it, but in fact there is disagreement (Tai; transl. by Sleeboom-Faulkner).

If it leads to the adoption of regulation (Lu; transl. by Sleeboom-Faulkner).

There were only two unconditional responses, albeit self-conscious ones:

> Of course, for all that we do is for the patients and we always listen to their views. I don't want to say that we are representative, however (Zhai; transl. by Sleeboom-Faulkner).

> No, because the international stem cell organisation already has stem cell research guidelines (Du).

Although, when pressed, stem cell scientists could think of reasons for holding public debate, in the first instance they regard debate as limited to the participation of intellectuals. Nevertheless, scientists could think of a limited number of ways in which public debate would include the 'masses' without threatening their work. This would not basically alter the national risk signature for hESR in China.

Some stem cell scientists, particularly internationally active principal investigators, have ideas about how debate could and should be organised in China. Some emphasise the role of the government. One leading scientist from Shanghai, Professor Jing, claimed that:

> Our government recently started a project to ask people from all walks of life about their attitude...They give a sum of money for researching people's views on this. They are still doing it (Jing; transl. by Sleeboom-Faulkner).

In Wuhan, a large industrial city in the inland Province of Hubei, this view seems entirely out of place. A leading scientist in Wuhan thinks that the government should create open discussion about oocyte donation, embryo donation, informed consent, the distinction between therapy and human experimentation, and other bioethical issues, but he does not think that general open debate is possible in China:

> Now only a few people decide things and make a discussion (Hu).

In even more isolated areas, a leading stem cell scientist/reproductive doctor expressed disillusionment about the ability of the government to organise bioethics discussions. He advised emphasising the training of physicians instead:

> Well, I think it is more important that the government encourages more general moral attitudes. I think it is even more important that medical schools emphasise moral education more (Hui; transl. by Sleeboom-Faulkner).

The reason for this call for government intervention is that Hu, together with many other scientists, believes that ordinary people's education and culture do not teach them how to make balanced bioethical judgements:

> It is not very easy. For there are many elements in Chinese traditional culture that are not very rational or ethical. In Chinese tradition, it is possible that the desire to have more children is emphasised over human rights. Chinese traditional culture is based on the household not on the individual (Hui; transl. by Sleeboom-Faulkner).

Bioethics aimed at autonomous individuals, according to this scientist, is not suitable for a rural environment, where the attitude towards the family household is crucial. Research into public attitudes toward SCR is also rare in metropolises. In China's main cities, which have flourishing SCR, one can find some debate about SCR among intellectuals, while in provincial cities scepticism even about the existence of political will to hold such debates prevails. In remote rural areas, such debate is thought to be unthinkable owing to the 'low' cultural and educational backgrounds and poverty. In such a situation, the set-up of bioethics institutions in SCR centres has occurred in the first instance with formal research regulation and without public debate. But as a movement of bioethics supported by international bioethics groups has gained more recognition, scientists' autonomy and views on bioethics are being challenged, and so is China's national risk signature of hESR.

Among scientists one can find a great variety of attitudes towards public discussion, but there is general agreement about the difficulties in communication between the press and bioethicists. While only a few stem cell scientists make an attempt to support the discussions on SCR, some take an active part in the discussion:

> If we are going to push the development of stem cell research, we need to have such debate. For the more people understand, the more support you get. In my case, it is so. This is why, if I have some time, I write small articles for the paper (Deng).

Others are optimistic about the role that scientists could play:

> Scientists should guide the discussion. Scientists should first discuss this with social scientists and then maybe get a deep understanding and then send it to the media and the newspaper in an article (Bing).

However, the majority of stem cell scientists, for reasons explained above, think that public discussion would currently be unhelpful. Although bioethicists are sometimes engaged to get political support and maintain good relationships with hospitals (Du), most stem cell scientists feel they are little understood by them and have little faith in their basic knowledge of science (Zhao, Du, Hu and Hui). One prominent scientist, who is not alone in having engaged in debate with non-scientists, finds discussion with bioethicists on therapeutic cloning a waste of money and time:

> If you talk to people and they never take your idea, it is a waste of time. I was involved in two ethics conferences, both in Shanghai. I was the only scientist involved. I gave two different presentations on co-operation between ethics and science. But each time, when the experts and lawyers asked questions, I understood that they understood nothing (Zhao).

The difficulties this scientist experiences seems to be linked to the difference in interpretation of bioethical issues at home and abroad:

> It is different when I talk about it in Europe. In the EU [European Union] committee, after I gave a similar speech, I got feedback. Real communication. That is why I think that in China maybe in the future we need some new people, perhaps some scientists, and they really need to co-operate with foreign experts, not only do the translation (Zhao).

China's bioethics debate among experts on SCR is not well co-ordinated. Zhao's experience raises the question of whether Chinese bioethicists are able to mediate between scientists and regulatory political organs to facilitate international collaborations. A prominent scientist from Guangzhou, Professor Pang, expresses this problem:

> On the bioethical level, we nearly have an exact copy of foreign bioethical regulation. It is as if we are so stupid. You guys are so lazy. Can't you set up your own standards? I would advice that scientists would get together and have a debate in China with the people and set up its own rules on the basis of its own ideas. So if you then criticise then I would defend China's bioethics to the end (Pang).

Although it is well-known that foreign or international guidelines are copied, Pang feels that the reason for copying is that China lacks the co-ordination and communication necessary to set up guidelines. Bioethicists, according to this scientist, try to prevent scientists from setting up the regulation conducive to their work, such as allowing embryo donation for compensation.

Public discussion is perceived as a risk in the eyes of many stem cell scientists. The main reason is that public debate could unmask practices that could tarnish the reputation of China's life sciences, as this is part of the national risk signature. Thus, even to elite laboratories that have robust institutional review boards (IRBs) and ethical supervision, public debate would risk negative publicity.

Their weary attitude towards public debate does not make scientists autocrats: owing to experiences with the press and bioethicists, who have queried the problematic nature of under-supervised bioethics institutions, they prefer to keep the process of institution-building in their own hands. And as there is no established co-ordination process in place to involve members of the public in the creation of regulation, they fear that bioethicists and politicians will hijack the process at their exclusion. Although it is important to many scientists to have the flexibility to have bioethics institutions and undermine them, for many other scientists it is crucial to have solid and internationally recognised bioethics institutions in place. In China, scientists have no choice but to co-operate with bioethicists as, since the establishment of the National Bioethics Association in 2008, bioethicists have irrevocably become involved in discussion on life science research regulation. The extent to which the public shall be involved is another question. China does not have a tradition of civil society, and it remains to be seen if even non-governmental organisations (NGOs) will be actively involved in regulatory activities.

Case 14: The Professionalisation of Bioethics and Impoverishment of Civil Engagement in South Korea[4]
(*with Seyoung Hwang*)

In China, the professionalisation of bioethics has meant a shift between scientists taking the initiative to set up bioethical review together with official regulators and government advisors to the central involvement of bioethicists, often with a philosophical or legal background, as the pushers of the regulatory process. This shift has meant a small extension of public involvement in the process to intellectuals and a hesitant official press. In South Korea, however, civil society groups have been collaborating with scientists in the establishment of regulation from day one. We shall see that professionalisation in South Korea has had the opposite effect of that in China of decreasing public involvement.

Since the late 1990s, South Korea has actively promoted the life sciences, in particular SCR. Developments of bioethics institutions in South Korea illustrate how, although the development of formal bioethics in South Korea was organised in a top-down fashion, it co-evolved through the activities and notions developed by civic groups and NGOs. This participatory form of creating regulation has gradually led to the absorption of civic players in government-led processes of constructing regulation through the professionalisation of bioethics. This trend can also be observed in other Asian countries, including Japan and China. Specific to South Korea, however, is the initial strong impetus of civic groups in organising bioethical activities.

In South Korea, bioethics legislation in the late 1990s began as restricted to parliamentary discourse. As in other countries, bioethics expertise was only just emerging, and its authority was barely recognised at that time. As such, the Korean establishment of bioethical governance should be characterised as a top-down approach. As was the case in mainland China and Taiwan, South Korean society had only just undergone radical reforms. In the 1990s, when a dramatic political regime change from dictatorship to a democratic state took place, the voices of scientific experts and civic organisations increasingly expressed views about political and social issues (Park, 2006; Bak, 2007). This was also the case when, following the birth of Dolly the sheep in 1997 and the derivation of human embryonic stem cells in 1998, South Korea took steps towards the regulation of bioethics governance. When ethical controversy ensued, members of religious groups and environmental NGOs began to call for the setting up of a national bioethics committee. But, until 2000, the main efforts made towards governance consisted of regulating the prohibition of human cloning. Mainly involving regulators and science experts, these efforts were aimed at facilitating biotechnological development.

This case study aims to explain some of the co-constructive processes of (both emergent) technocracy and democracy in South Korea in the context of bioethical regulation of hESR. The study developed ethnographical methods for investigating the enquiry, including interviews with individuals who were participants in the bioethics campaign (1997–2005) with a variety of political and professional backgrounds, and stem cell scientists and bioethics experts whose work directly concerned bioethical governance. The fieldwork was conducted from April to December 2009. Interview data were interpreted by triangulating them with documentary sources already available, such as self-reflective accounts on bioethics campaigns written by interviewees, academic papers and media articles that directly addressed the activities in which interviewees were engaged, NGO websites, campaigners and research institutes, bioethics policy documents and so on. Such a variety of data sources provided useful tools with which to develop and examine concepts and categories such as membership, alliance, collaboration, network and so on by considering them as contested constructs that comprise the multiple interpretations of 'identity'.

The development of NGOs included feminist and environmentalist groups. In 1997, the Centre for Democracy in Science and Technology (CDST) was set up by civilians concerned about the role of science and technology in Korean society (CDST, 1999). It played an important role in propagating the need for discussion on bioethical issues. It was instrumental in creating the Alliance for Biosafety and Bioethics (established in 1998), and played an important role in the Urgent Cooperation Campaigning for Bioethics Legislation (established 2001). The CDST exerted much influence on the development of bioethics in South Korea by mobilising large-scale campaigns through alliances with NGOs, religious groups and a handful of experts. It also sought democratic forms of public engagement; for example, consensus conferences were organised in co-operation with UNESCO (United Nations Educational, Scientific and Cultural Organization) on various themes, ranging from genetically-modified food, human cloning and nuclear power. The CDST also proposed the Bioscience, Human Right and Ethics Bill in the form of a petition in 2000 as the alternative to the previous bills.

Since 2000, when media reports appeared on the derivation of stem cells from in vitro fertilisation embryos and therapeutic cloning, calls for regulation from NGOs and the Roman Catholic Church became louder. The legislation of bioethics became the turf war between pro-regulatory and pro-research groups, and between the Ministry of Health and Welfare and the Ministry of Science and Technology (MoST), which represented each group's interests respectively. An inter-ministerial office, the Bioethics Advisory Committee, was established to initiate co-ordinative discussions. The establishment of this committee, composed of 20 members, including humanities scholars, NGO representatives, biotechnology professors, a Catholic priest and medical doctors, led to the establishment of the National Bioethics Association (NBA) under the MoST in 2000. The NBA included social scientists, NGOs, religious leaders, bioscientists and medical scientists, and largely created the basic framework for the Bioethics Law in May 2001. But once bioethical concerns became official political issues, they attracted public interest and led to conflict and the polarisation of views between what were portrayed by the press as 'science' and 'ethics' supporters.

A notable conflict developed between the Catholic churches, which opposed therapeutic cloning and embryo research, and scientists, who criticised any obstruction to embryo research. The strong reaction of scientists led the MoST to declare that the NBA's position did not represent the official stance. Subsequent conflicts were framed in the mainstream media, often typified as 'scientific advance fettered by ethics' (Kim, 2004). The success earned by stem cell scientists in Hwang Woo-Suk's team, and his funding, augmented the growing patriotic and optimistic atmosphere among the public, culminating in his international publication successes in 2004 and 2005. In 2005, the Bioethics Biosafety Act created more stringent conditions for somatic cell nuclear transfer (SCNT) research, but Hwang Woo-Suk was able to take advantage of interim measures to apply for research permission for patient-specific embryo cloning. Nevertheless, even before the 'scandal' occurred, many critical voices on 'oocyte donation' and 'authorship' had expressed their protest against unethical research practices (Cyranoski, 2004; KBA, 2004).

But it was the revelation that Hwang's publications were based on scientific fabrications rather than ethical problems that lost him most of his public support and research funding (Kim, 2008a). Although the role of civic players in the development of regulation had been considerable, new regulation was now built on a more formal bioethical basis. Thus, the 2008 amendment of the Bioethics and Biosafety Act was aimed at bureaucratic control of research oversight. But, focused on clearing up the deficiencies in the oversight directly concerned with the Hwang scandal, criticism continued, aimed at inconsistencies in the oversight. For example, rules were specified allowing the donation of eggs for SCR, but they did not need to be met in the case of egg donation for infertility purposes.

Now that bioethics had become a matter of international prestige and a requirement for international high-quality science publications and international collaboration, bioethical regulation became a matter of top-down implementation. In short, owing to international incentives, bioethics activism underwent professionalisation and became established within the academic community. This change in the power configuration meant that grassroots campaigners, such as CDST, could no longer play a role in public discussion through the formation of alliances with concerned scientists. Although the extent of the role of these civic players is hard to measure, and although it has been argued that their effect on science regulation, especially since 2009, has been minimal (Kim, 2008b), it is clear that in South Korea great effort has been exerted by civic campaigners to pressure official parties to engage the public in a democratic manner. The role of scientists in the public debate was important during the process of bioethical legislation. But in the aftermath of the Hwang Woo-Suk scandal, the Korean scientific community has been much criticised for its lack of self-discipline. It has been suggested that stem cell researchers would have preferred soft guidelines rather than a legal framework for bioethical regulation to keep external intervention to a minimum. Nevertheless, it is also clear that they have increasingly accepted that the stricter policy approach adopted after the Hwang scandal was inevitable in order to restore public confidence in the governance of SCR. In terms of their notion of public discussion of ethical issues, scientists regarded more accurate, science-based communication as necessary for moving away from either 'blind' support for or objections against SCR.

Case 15: HESR and iPS in Japan: From 'Unethical' Disentanglement to 'Reckless' Application[5]

Bioethical regulation of hESR in the UK has been cited as rooted in public opinion, exemplified by the engagement of the public in debates on embryo and oocyte donation (http://www.hfea.gov.uk/consultations/index.jsp). Developments in the Japanese governance of hESR show how regulation can be rooted in society differently. Although its public engagement strategy received little public response (Kato, 2005), Japan's stem cell regulation was, nevertheless, largely steered by society. For, although the state adopted a permissive political stance towards hESR, its regulatory practice largely relied heavily on the opinions of particular social groups. But the regulation was reviewed in 2007, after Yamanaka's work on iPS gained fame among the public. The 2007 reappraisal of hESR in the light of iPS and the efforts made to advance to clinical applications in 2008 clearly reflect a change in the orientation of policy-makers toward regulation. To understand Japan's policy turnaround, we need to observe how ethical science is weighted against, or uncritically replaced by, risky science. This case study shows how the regulatory emphasis on what is regarded as ethical in the case of hESR was confused with a 'safe' procedure, and how a change in favour of 'ethical' iPS according to stem cell scientists actually formed the reckless option.

This study draws on data gathered during ten weeks of fieldwork in 2006 and six weeks in 2008 to gain insight into valuation of reproductive materials and potential scientific discoveries. The fieldwork consisted of archival research and semi-structured interviews with 40 stem cell scientists, 25 policy-makers, two students, eight housewives and six religious professionals. The three latter groups were chosen to gain insight into the views of social groups, whose rough equivalence in Europe and the USA object to hESR. Data obtained from interviews with stem cell scientists pertain to their views of regulation, its implementation and public discussion, while data from interviews with students, housewives and religious professionals served to obtain insight into popular understandings and support for embryo research and iPS. Interviews with policy-makers yielded insight into how various regulators and committee members deal with the diversity of views among the public and scientists in regard to hESR regulation and its implementation. The data obtained made me aware that it was the fear of scandal and a tarnished reputation rather than the generally controversial nature of hESR that was problematic in the regulation of hESR in Japan. Pseudonyms have been used so as not to attract attention to individuals, although none of the interviewees, when asked, expressed objections to being cited. Interview questions concerned knowledge about, and the ethical issues of, regenerative medicine. In the case of scientists, they also included questions about the regulation of the research and work in the laboratory. Interviews took place mostly in the Kansai and the Kanto regions, where the main SCR hubs are located, and interviewees were approached through contact networks, starting at the twelfth meeting for Genetic Engineering and Regenerative Medicine on 22 April 2006 in Kyoto.

In contrast to the impressions evoked by broad claims of the unity of the Japanese stance on bioethics (Robertson, 2005), Japanese views on the embryo, gametes and the fetus vary greatly. Interviews show that views on the embryo range from regarding the embryo as the 'germ of life' and the seed of life to the embryo being sacred among Japanese sects. Views on the treatment of gametes include no objection to the donation of oocytes, a forbidding attitude to oocyte donation, and the attribution of *kokoro*, or mind, to ova (see Sleeboom-Faulkner, 2010d). Although many Japanese people, in particular scientists, say that fetuses have no spirit, many women partake in ceremonies for deceased fetuses (*Mizuko kuyo*) (LaFleur, 1992). But it is not entirely clear to what extent this means that women regard the fetus as having a spirit (Sleeboom-Faulkner, 2010c).

Nevertheless, the public discussion on hESR in Japan has been limited (Kato 2005). Only the views of a few groups, such as the anti-eugenics network (which broke up in 2007), the *Soshiren* (Japanese women's movement), some patient movements and religious groups have been voiced in the media. Multiple positions on hESR co-exist varying from support for SCR to help handicapped people and for infertility treatment, opposition to SCR when regarded as too artificial [*fushizen* (unnatural)] and too expensive, opposition to abortion and the inter-human transfer of biomaterials for religious reasons, and opposition to hESR as a protest against the donation of reproductive materials by women (see Sleeboom-Faulkner, 2008b; Kato and Sleeboom-Faulkner, 2011). But even though ample disagreement exists about hESR in Japanese society, and even though the government has invested heav-ily in national debate on the subject, public debate remained low-key.

The low-key nature of the debate, according to group discussions held in December 2008, was owing to the general aversion among Japanese people to 'unnatural' practices, and 'high-tech' and 'commercial' atti-tudes towards life—a view also voiced among some members of disabled movements (interview Chiba, 1 June 2006). Although approximately 30% of the interviewees viewed human fertilised eggs as 'human lives', over 40% said that they would approve their use in research conducted to advance medical science. What is regarded as problematic by most, however, is not so much the principle of using embryos or oocytes in research, but the idea of mainly focusing on high-tech methods for extending the lives of the ailing elderly, rather than improving the qual-ity of life during the 'natural' life-span.[6] Another relevant point is that approximately half of the scientists (21) and nearly all non-scientists (25) said that the donation of reproductive materials should take place only to help kin. Potential donors also speak of the motive for embryo donation in terms of kin. Thus, research shows that couples asked to donate embryos usually respond in terms of 'motherhood ethics' justify-ing the answer in terms of what is 'best for the child' (Kato and Sleeboom-Faulkner, 2011). The commercial incentives and the impersonal nature of the research were most mentioned as the reasons for the aversion [*kirai* (dislike); *iya* (disgust); *fushizen* (unnatural); *ayashii* (dubious, suspicious)] felt against the use of reproductive materials in scientific applications. Central here then is the controversial nature of the disentanglement (Callon, 1998) of reproductive materials from the intimate sphere and its insertion into a commercial and impersonal economy.

When Yamanaka Shinya and Kazutoshi Takahashi published their research on iPS cells derived from mice in 2006, not many Japanese people had heard of stem cells and pluripotency. In 2006, interviewees revealed that most laypeople had not heard of stem cells, although some could relate the subject to the Hwang Woo-Suk scandal surrounding the fabrication of research data and unethical oocyte donation in South Korea (Sleeboom-Faulkner, 2008b). But according to a survey conducted by Shineha et al. (2010), in 2008 more than 80% of Japanese people had an opinion on regenerative medicine and expressed understanding of the significance of 'pluripotency'. Similarly, interviews I conducted indicated an almost universal awareness of the meaning of iPS and pluripotency among housewives and students.

iPS cells are products of the direct reprogramming of somatic cells to an embryonic-like state. This process involved the introduction of a limited set of transcription factors (initially four) and feeder culture (under embryonic stem cell conditions) to 'fool' the somatic cell into reprogramming. Although Yamanaka and Takahashi, in October 2006, were the first to 'discover' iPS, when they tried to apply their technique to human somatic cells, they had competition from Jaenisch's team at the Massachusetts Institute of Technology, who published their laboratory's research results in the same month (November 2007).

IPS in Japan was received with great enthusiasm. It was front-page news for weeks, and was discussed in all newspapers, on TV and on the Internet. IPS was advertised as a bioethical cure for old-age diseases, not requiring oocytes or embryos. Compared to SCNT, iPS promised to produce pluripotent cells with the same traits as human embryonic stem cells: pluripotency was the buzzword in the media. iPS was portrayed as an endless resource for producing healthy cells that could be used for clinical application. Also, in other respects, iPS cells compared favourably to other cells, as they occur in humans naturally (Nishikawa et al., 2008). The only problem—already on the way to being solved—was their tendency to produce teratomas, and the use of a viral vector. Using fewer and other genetic factors, and finding an alternative to viral vectors, was to solve these problems.

Much was made of Yamanaka as the 'good' scientist, who persevered in his quest to generate new medicines, spinning out Yamanaka's narrative that once it had occurred to him that the embryo he saw in a friend's laboratory through a microscope could have grown into a human like one of his daughters (interview 6 May 2006). When there was talk of the clinical application of iPS, the press quoted Yamanaka as saying that the simplicity of iPS warranted regulation, as in the near future anyone would be able to create them, including iPS cells that could be developed into gametes (Alford, 2008; Anon, 2008). The speed with which government support was mobilised for the new scientific discovery was unprecedented (interview Sato 17 March 2008). Prime Minister Fukuda, then the chair of the Cabinet's Council for Science and

Technology Policy (*Kagaku Sogo Kenkyukai*), ordered an urgent policy review, and the Ministry of Education, Culture, Sports, Science and Technology (MEXT) immediately started developing policies to regulate iPS and its future applications. Meetings with civil servants and scientists were promptly organised the month after Yamanaka and Takahashi's article was published in *Cell*, and by 22 December 2007 MEXT had made plans to raise the funding on iPS research from 270 million yen (US$2.5 million) for research in 2007 to 2.2 billion yen for the 2008 fiscal year, pledging 10 billion yen over the next five years (Cyranoski, 2008). Half of the 2.2 billion yen was to be pumped into Yamanaka's iPS research to be housed in a new institute for iPS at Kyoto University in the new national centre for Integrated Cell-Material Sciences (iCeMS). Furthermore, the Ministry of Health, Labor and Welfare would channel close to 100 million yen in the 2008 fiscal year directly to Dr Yamanaka, in addition to 410 million yen for regenerative medicine infrastructure, such as a cell-processing centre (Cyranoski, 2008).

This move was made not just because research on pluripotency was no longer to be obstructed by bioethical hurdles, but also because many scientists, including Nakauchi Hiromitsu, Chairman of the Japanese Society for Regenerative Medicine, had pointed out that other countries could win the competition for medical applications and patents (see http://www.jsrm.jp/index.html). This concern was expressed in the regulation of the Japan Science and Technology Agency, which emphasised the need to pay attention to competition, the nature of patent acquisition and bioethical appropriateness (JSTA, 2009).

But the majority of interviewed scientists in various fields of SCR (30; 75%), when asked for their view on current policies on the allocation of science resources, doubted the wisdom of the government's heavy investment in iPS. They argued that if one wants to cure diseases such as Alzheimer's or Parkinson's, a 'rational' approach requires one to try out various possibilities: using different methods and different cells; comparing the success rates of fetal cells; and hESR through SCNT and iPS. At present, they argue, it is not yet clear if any cell therapy works. And, if it does, it is not know which is most effective and which is safest. However, the situation in Japan was not conducive to such diversified approach.

The disagreement in Japan about priorities in regenerative medicine is apparent in the development of bioethical regulation. An officially permissive approach permitted hESR in studies aiming to cure serious diseases, guiding the creation of human embryonic stem cell lines and 'special embryos' (*toku teihai*), while the urgency of regulating 'therapeutic cloning' and iPS in advancing research was recognised. But this permissive attitude was overshadowed by a bureaucratic approach to hESR. Complex procedures for both establishing embryonic stem cell lines and their use involving double-review by IRBs and MEXT could take longer than a year (Nakatsuji, 2007), while changes in protocol or research team required additional applications. Special equipment was required for storing and discarding human embryonic stem cells, for example a furnace, and a room exclusively used for human embryonic stem cells. Thus, human embryonic stem cell lines could not be stored in the same room as animal cells, even though they were maintained by an animal (mouse) feeder. Ordinary human embryonic stem cells were to be respected, while their supposedly identical iPS counterparts did not receive any such treatment. Thus, the permissive approach aimed at advancing hESR was countered by a purist approach adopted by the regulatory bureaucracy, the accountability for ethical research of which makes it responsible for improprieties and scandal.

The bureaucratic approach is informed by a circumspective attitude towards propriety. According to cell banker Murakami, civil servants run the risk of losing their jobs or of being sued if they are seen to be inappropriately backing fetal SCR or gene therapy, or by allowing hESR without the utmost scrutiny. Gene therapy trials in Japan, according to fetal scientist Kawahara, are usually only attempted when US and European research has shown it is 'safe'. Guidelines issued in the spring of 2006 stipulated that oocyte donation informed consent procedures are only put in place when primate SCNT has been shown to be possible, or when human SCNT has succeeded abroad. These conditions took into account the increasingly louder views of feminist and disabled movements who stand up for the position of women and paternalistic attitudes associated with Japanese doctors, and a growing distrust among the public of scientists' 'Western' and 'rational' approaches towards medicine, which are blamed for losing the 'Japanese' human factor essential to a society plagued by suicides and an increase in 'un-Japanese' practices such as 'brain death', 'euthanasia' and prenatal testing (Morioka, 1995; Lock, 2002; Tsuge, 2010).

Reliance on compromises between permissive and purist approaches created friction among policy-makers, scientists and social movements, which was lessened by the discovery of iPS. The enormous attention on iPS in late 2007 resulted from the relief of not having to deal with the thorny bioethical issues around embryo research in a field in which Japan had taken the global lead. The media speculated on the possibilities of using iPS cells in regenerative therapies without the problems of immune reaction, and ethical dilemmas associated with oocyte and embryo donation. In December 2007, it was decided to invest substantially in iPS development and therapies, mainly institutionalised in the Centre for iPS Research Applications and iCeMS in Kyoto University, and to speed up the development of iPS and patent law concentrated in Kyoto. At the beginning of 2008, the creation of so-called '*Suupaa Tokku*'—special research fields, such as for iPS and hESR, organised around licensed research institutions—was designed to speed up the review of research proposals and the development of protocol (see Kitagawa and Woolgar, 2008). A special task force from MEXT allowed therapeutic cloning in December 2007 (MEXT, 2008), and in October 2008 various scientists expressed the belief that rules regarding oocyte donation would become less stringent in the near future. In October 2008, it was decided to loosen regulation around the application for hES lines (regulated by the MEXT in 2009[7]), and in November 2008 the ethics committee of the Cabinet's Council for Science and Technology Policy (CSTP, Kagaku Sogo Kenkyukai) allowed the creation of gametes from pluripotent cells.

Contrary to expectations, the majority of interviewed scientists regarded as 'reckless' (*musekinin*) the radical switch in research emphasis from a purist and strictly regulated hESR to bioethically correct, but unregulated, iPS research in late 2007.[8] Pluripotent cells created from iPS cells were presented as identical to human embryonic stem cells, even though scientists in Japan and in the USA commented on the limited available knowledge about the molecular processes underlying the reprogramming of somatic cells involving iPS. Nor had iPS cells been directed to differentiate into specific functional tissues or organs. And although alternatives to the use of teratoma-generating retroviruses were being sought, the introduction of genetic factors still meant that genetic alteration could take place, the consequences of which were hard to gauge. Thus, there was not yet confirmation of a genetic identity between the cell type generated from iPS and that of the patient. And, according to all interviewed stem cell scientists, the behaviour of iPS cells and their therapeutic effectiveness depends on the environment in which the cells are going to end up. Concentrating on iPS at the expense of other branches of SCR, according to most scientists, was a risky strategy: it threatened the loss of current expertise and it was based on unexplored presumptions about the nature of iPS.

The need for disentangling biomaterial from potential human life to support the artificial extension of ageing life in Japan has been a topic of critical public discussion about its naturalness and ethicality. As a result, hESR regulation was influenced through group notions of ethicality, while social hype and government ambition in the case of iPS led to relative insensitivity to the risks associated with planned application in a similar research field. A majority of interviewed stem cell researchers indicated that the resultant 'recklessness' was equally damaging to scientific advancement. This case study shows how the notion of 'ethical' research is very different from 'safe' research. In the case of Japan, public discussion and public views were necessary to define the ethicality of hESR. The perceived experience that such 'ethical' regulation was responsible for hampering science research made a shift to investing into iPS seem a much 'safer' option. Potential iPS applications became associated with the possibility of wide commercial distribution and large-scale application in regenerative medicine. And in 2008 there was little sign that iPS research would receive the same scrutiny as hESR had been subject to, even though it was accompanied by short-term plans for risky clinical application. The decision to opt for an approach unhampered by public views on the meaning-of-life issues (iPS), then, had taken place at the cost of considerations of risk.

Conclusion

The life assemblages in East Asia discussed in this chapter show major differences between the attitudes of scientists towards public discussion of hESR. Thus, scientists in China feared that public discussion would unnecessarily disadvantage hESR in China and influence its national risk signature in a way that would not work for them. But scientists in Japan had no such scruples about involving the public in ethical discussions on their work; they regarded public discussion as an asset to hESR in Japan. In fact, a strong belief can be found among scientists that, if only the public would listen and be involved in debate, their understanding would influence the tight regulation favourably. In a way, this attitude benefited them, as the media hype around iPS garnered the support of the public and prompted the government to invest in the field and to start altering regulation for SCR. Ironically, it also caused anxiety and consternation among scientists about pushing research in one direction without comparing the safety of various science methods first. The high sensitivity of Japanese society to scandal, some scientists argued, could turn any scientific misconduct or recklessness regarding iPS into a major social risk.

In China, Japan and South Korea public discussion has taken various forms, varying from debate on the Internet, debate represented in medical textbooks, and discussion among bioethics committees, scientists, medical professionals and civil society activists. In China, most scientists regarded the stimulation of discussion by 'the people' as threatening the field to various extents; in Japan, public discussion was welcomed. And, in South Korea, scientists' views on engaging the public were strongly divided. In the three countries the professionalisation of bioethics had a different effect. In China, it increased the, for a long time, much limited discussion, while in South Korea and Japan it actually intensified the role of professionals, exhausting competitive voices from civil society initiatives. Thus, the Anti-Eugenics Thought Network had to give up as a result of the high pressure, time and costs associated with activism.

Japan's government has initially been careful to consult multiple stakeholders when regulating hESR. It arranged for public discussion to be held, and regulatory authorities included minority views, which were included in bioethical guidelines. Although safety was a major issue for scientists, ethical issues related to the role of women and the moral status of the embryo were pronounced. The embryo has to be respected, and human embryonic stem cell lines were not to be stored with animal lines. There was a clear dividing line between the ethical and risk issues visible. But when 'ethical' iPS entered the scene, the sense of difference was watered down. IPS was fully supported, and when it became clear that advances in iPS research would profit from comparison with human embryonic stem cells findings, the bureaucracy around hESR was diminished in 2009. This development differs from that in China, where ethical issues were more important for the recognition of the validity of hESR than for the ethics itself. Only a few scientists made a clear distinction between 'ethical' and 'good' science applications. This is not because the embryo has no moral status in China; it is because the voices that could potentially express their view on its moral status were not systematically included. With the professionalisation of bioethics new formulations are being created that co-ordinate selected views on the moral status of the embryo with regulation conducive to 'ethical' SCR. Whether this trend is acceptable in the context of the various life assemblages remains to be seen.

7
Life Assemblages and Bionetworking: Developments in Experimental Stem Cell Therapies in India and Japan

Having discussed discourses around bioethics and stem cell research and notions of risk in China and Japan in Chapters 5 and 6, this chapter focuses on the ways in which stem cell research has developed in conjunction with the provision of stem cell therapy services in India and Japan. The enterprise of clinical stem cell science constitutes a complex social, medical, economic, regulatory and scientific venture that resonates across local and national borders. In contrast with research into stem cells, the provision of stem cell therapy is predominantly market-driven, involving a struggle among players at various institutional levels, including the state, public and private healthcare providers, and regulatory bodies (Salter et al., 2007; Salter, 2008). High expectations are raised about stem cell transplantation as the hopeful answer to a large number of acute and chronic disease conditions for which modern and conventional medicines have little to offer. These stem cell transplants mainly take place in private sector research-cum-hospital set-ups, although some public or state hospitals have also ventured into this field. In recent years, India has emerged as a new global player in this field. Although the evidence of the efficacy of adult stem cells in curing a wide range of disease conditions is questionable, many Indian and Chinese healthcare centres have been carrying out such therapies as experimental research or experimental therapy.

This chapter shows how characteristics of life assemblages are crucial to the way that patients are recruited and research therapies are organised transnationally. In the links between life assemblages, forms of 'science collaboration' play a crucial role. What we call bionetworking activities underpin the promotion of stem cell therapies and research, and form the basis of research conducted in this area of cutting-edge life science competition. Although science collaboration aims to pool together resources for the purpose of research that yields scientific knowledge or applications, most scientific research also serves other purposes, such as gaining fame, acquiring patents and attracting funding. In what we call bionetworking, non-scientific activities underpin the scientific ones. To understand how and why collaboration between researchers, therapists and other actors occurs across cultures and countries it is essential to have an understanding of the various life assemblages involved.

The kind of collaborations explored in the case studies discussed here advertises stem cell research and therapy provision activities as bona fide, while profiting from questionable experimental stem cell applications. Bionetworks here link key individuals involved in research and healthcare organisations that take advantage of the unequal socio-economic contexts in which research takes place and healthcare is provided. The scope for bionetworking activities increases under conditions of health inequality, part of what Petryna (2007) calls the remake of the global geography of human experimentation. Bionetworks constitute a plurality of actors forming biotechnical ventures (Waldby and Mitchell, 2006) that work across geographical space, forming what Sunder Rajan (2007) calls 'biocapital'. To sustain its infrastructure and research, bionetworking makes use of capital from foreign investment, know-how and therapy fees. Costs depend on the institutional and legal structures within which these companies operate. The production costs of experimental therapy provided in India and China can be relatively low, while investment into the therapy's infrastructure, such as legal permissions, hospital facilities, quality certificates, advertising and intellectual property rights (IPR), can be kept to a minimum through bionetworking activities.

International Collaboration

The concept of international collaboration describes both a market mechanism and a discourse. As a conceptual tool for describing markets, it indicates the existence of a mechanism that harmonises disparate interests with the unequal distribution of surplus values: for, exchanges in international collaborations are not so much based on the ideal of benefit sharing between various exchanging partners as they are aimed at obtaining the potentially most profitable exchange among all possible exchanges within reach in the face of global competition and competitors at home. Although mutual benefit-sharing is often the stated aim of international collaborations in the life sciences, they can hide unequal relations, which enables exchanges on the terms set by those dominant in the field. Using the life sciences and internationally accepted bioethics research regulation as a template, official regulatory discourse, including bioethics regulation, masks what are, in fact, messy processes of market exchange. These points have also been argued by Kaushik Sunder Rajan (2007) and Adriana Petryna (2005, 2009), who claim that unequal conditions of life in low- and middle-income countries are the groundwork upon which harmonised markets take shape. In the case of the subject recruitment for clinical trials, bioethics regulation may cover up the circumstances under which humans are experimented upon (Sunder Rajan, 2006: 46; Petryna, 2009; Sleeboom-Faulkner and Patra, 2011).

But this analysis can be taken a step further. Bioethics discourses, besides enabling exploitative exchanges, can also conceal research practices. For although bioethical regulation is required for conducting internationally acknowledged research and publishable papers in peer-reviewed academic journals, the establishment of internationally acknowledged bioethics institutions and the cultivation of bioethical values could make potential collaborators less attractive. To overcome this obstacle, collaborations based on harmonised bioethical regulation must, in fact, mask the malleability of bioethical values in research. Such bionetworking activities include the groundwork needed in the experimentation with a bioethics discourse flexible enough to enable both collaboration and hidden practices without overtly harming the interests of collaborators involved. The honing of social and language skills, the establishment of regulation and the creation of favourable conditions aimed at enabling potential collaborative partners usually, in fact, occur long before the collaboration itself takes place–if it takes place at all. The active involvement of scientists in such groundwork is evidence that stem cell scientists in low- and middle-income countries are not just passive objects forced to adopt what they regard as alien Western guidelines for bioethics; a majority, in fact, actively tries to fit into this international field of science production, creating and shaping new science institutions and cultures. Here it is not so much the 'ethical variability' that facilitates transnational science research, but the active manipulation and remoulding of standardised bioethics that enables and encourages 'competitive' research practices. The question arises, then, of how bionetworking remoulds bioethical values and which values are neglected in order for scientists in low- and middle-income countries such as India and China to harmonise their laboratories with 'international' bioethical expectations and make them preferential as potential collaborators at the same time.

Bionetworking operates across local, national and global levels, promoting and sustaining a bio-political and bio-economic enterprise, also in the face of incapacitated state power and differences in health provision, access and purchasing power. It proceeds from the management of complex inter-relations and lopsided interdependencies between stakeholder groups, such as patients and their relatives, researchers and physicians, regulators and the media. Two case studies illustrate the ways in which bionetworking proceeds in experimental stem cell practices. Case 16, on the promotion of experimental stem cell therapies in India, shows how networking at a local level interlinks various therapy-providing institutions, while Case 17, on bionetworking between Japan and India, illustrates the transnational workings of growing therapy-cum-research collaborations across borders. In short, this chapter explores how international collaborations in experimental somatic stem cell therapies take advantage of international and local inequalities, and how science management actively creates and shapes institutional structures that fit the purposes of scientific advancement, fame and profit. It investigates how the global moral economy facilitates thriving experimental stem cell research and stem cell therapies in the embedding of life assemblages in India and Japan.

Case 16: Promoting Experimental Stem Cell Therapies in India—Bionetworking for Patients[1]
(*with Prasanna Kumar Patra*)

Over the last decade India has emerged as one of the preferred locations for stem cell-related research and as a site for experimental stem cell-based intervention activities. It has become possible owing to, on the one hand, proportionally greater investment by the state in medical biotechnology research and development (Salter et al., 2007), a state of stem cell 'governance vacuum' (Sleeboom-Faulkner and Patra, 2008) and a lack of authority for regulators to enforce policies (Pandya, 2008). On the other hand, India has benefited from the policy decisions and ethical dilemmas over embryonic stem cell research in the USA (e.g. the National Institutes of Health's moratorium on public funding for stem cell research in 2001) and stringent regulation on the clinical application of adult stem cell therapies in the 'West' (Kulkarni, 2008; Salter 2008). Government policies on stem cell science in India have been very supportive of programmes that aim to promote both basic and translational research in view of its potential application (Sharma, 2006). Among policy issues, public–private partnership in basic and translational research in stem cell science is given importance, which signifies an atmosphere of renewed vigour, flow of capital and strengthening of capabilities. There are many such initiatives in India among academic institutes, hospitals and industries in the field of stem cell research and therapy.

In India, stem cell applications or therapy services are on the way to becoming a common practice—with many tertiary-level hospitals and healthcare centres entering into stem cell research and/or clinical application as part of their service provision. Currently, stem cell applications in India are undertaken for a variety of disease conditions, such as spinal cord injury, muscular dystrophy, liver cirrhosis, diabetes, Parkinson's disease, ischaemic heart disease and retinal pigmentosa. The sources of stem cells vary from centre to centre. They usually derive from bone marrow, cornea, liver, peripheral blood, umbilical cord, fetal stem cells and embryonic stem cells of animal origin (Shroff, 2005), but here the discussion surrounds adult stem cell therapies only. In general, all human stem cell research and therapies require official registration and ethical clearance. Experimental stem cell therapy and research are mainly provided by private hospitals, as well as a few public ones, such as the All India Institute of Medical Sciences (AIIMS). As the nodal agency, AIIMS reported conducting a multi-centric clinical trial, using stem cell therapy, organised at five centres across India, for diseases including myocardial infarction, cardiomyopathy, muscular dystrophy, cerebral palsy, diabetes, retinal pigmentosa, spinal cord injury and amyotrophic lateral sclerosis. These trials started in 2003 and more than 750 patients have already participated in the trials (*The Hindustan Times*, 2006).[2]

Clinical applications of stem cells in treating 'untreatable' diseases take place mainly in private sector hospitals. Such hospitals use autologous adult stem cells (where cells are removed, stored and later given back to the same person), which are claimed to be safe and ethical therapy modules. Two issues are of concern here: the questionable scientific basis, medical efficacy and regulatory permissions for using adult autologous stem cell therapies, and the way adult stem cell therapy is being promoted ambiguously as experimental therapy in many service-providing centres. As interviews with scientists indicate, there is a widespread belief that adult autologous stem cell therapy, unlike embryonic stem cell therapy, is free from ethical problems and has a great therapeutic potential (Khan, 2007). But, as shown by the scientific literature (Strauer and Kornowski 2003; Passier et al., 2008), more basic research is needed on its efficacy and side effects before such claims would be accepted in more strictly monitored environments. In this context, the use of the concept of experimental research is premature.

Service Providers: Physicians and Co-ordinators

In our analysis, service providers are physicians, including individual practitioners associated in hospitals, who provide stem cell therapy in India. Although there is a great variety in the kind of hospitals that offer stem cell therapies, there is a pattern that shows systematic links between large well-known hospitals and small hospitals that engage in underground collaborations. Bionetworking plays a crucial role in the co-ordination of their scientific activities. The reasons provided for engaging in stem cell therapy vary. Physicians who provide it as 'supportive' therapy along with their conventional treatment modules express the belief that stem cells provide better results. Other independent physicians connected with researchers and physicians in the country and abroad work in tandem on stem cell research. Circumstantial evidence and participant observation at some field sites in India show that some work as facilitators for clinical trials and experimental stem cell therapies. Although some physicians in smaller towns/cities provide experimental stem cell therapy to earn quick money, others insist that their treatment works and defend the moral correctness of helping patients by providing them with experimental stem cell therapy.

It is not clear how many centres or hospitals in India provide stem cell therapy as a regular medical practice, as a number operate underground, in addition to the many centres that provide stem cell therapy openly. But there is a general pattern where big private hospitals or healthcare centres function as a 'hub' and are known to the world, while small, less obtrusive hospitals–the 'spokes'–are not. Every other day in the media physicians report breakthroughs in stem cell therapies for 'untreatable' diseases made in both public and private sector hospitals. The individual statements of physicians, however, cannot be understood separately from the nature of the clinics that they work for or lead, many of which are 'spokes' connected to the hub of a large network. The clinics providing experimental stem cell therapy have distinct characteristics. First, they are usually individually driven and network-based. The stem cell division of the centre revolves around a key figure or an influential individual physician who has a wider network across local, national and global levels. As the therapies are over-hyped and their clinical applications do not require highly skilful medical expertise, these centres manage with a few physicians and paramedical staff. Second, mutual interests are promoted through public–private partnerships. As research and development (R&D) in the field of science and technology in India was, until recently, monopolised by the public sector, the reputation of which is superior to private enterprises and which receives considerable public funding, emerging private clinics and hospitals are attracted to the public sector for social recognition and acceptance. Thus, in the field of stem cell therapy, private hospitals (hubs) enter into collaboration with reputed public institutions to gain public recognition. One example is the recently held First International Stem Cell Summit at Chennai in joint collaboration between a premier public sector technological institute and X Institute for Regenerative Medicine (XIRM), which also collaborate on other aspects of stem cell research. But as funding for R&D has lately increased in the private sector, public sector centres are also motivated to collaborate with private sector enterprises.

Two Case Studies: Institutional Complexities

The manifold links between health institutions and the diversity of collaborative forms between public and private elements of these healthcare centres make for opaque organisational structures. This opaqueness forms an important feature of the bionetworking strategies in experimental therapy provision illustrated by a description of two fieldwork settings.

A. Setting 1, Chennai: XIRM

The medical centre named XIRM, in Chennai (Madras), capital of the state of Tamil Nadu, is emerging as one of the healthcare hubs in India. Many of its corporate medical centres have started to focus on biomedical education, research and translational research. These centres provide 'super-specialty medical facilities', and are popular sites for global clinical trials. In Chennai, there are 14 or more institutes working on basic research and clinical applications of stem cells (Som, 2007). The X Group of Hospitals—a private enterprise and one of the largest referral centres in Chennai—has increased its strength 20-fold (from 25 beds in 1997 to 500 beds in 2008) since its establishment in 1997. A 2006 report describes how the X Group has acquired 15 clinics in 13 healthcare locations across the state of Tamil Nadu through their 'hub and spokes model'. Each of the clinics has one or two consultants (depending on the size of the clinic), a pharmacy, and a laboratory with electrocardiography and X-ray facilities (Rangarajan, 2006). XIRM is part of the X Group of Hospitals, and focuses on stem research and therapy. XIRM claims to provide adult stem cell therapy for several medical conditions, including spinal cord injury, liver cirrhosis, Alzheimer's disease, cardiac infarction and cerebral palsy. XIRM also claims to receive collaborative inputs from international centres such as the Biotechnology Department of Temple University and the Kennedy Kreiger Institute (both in the USA), Nottingham University in the UK, and the Indian Institute of Technology, Madras and the University of Madras in India (ISCSI, 2008).

Prior to the formation of XIRM in January 2007, X Hospital had collaborations with Z Centre for Regenerative Medicine (ZCRM) in Chennai. ZCRM provided technical services, such as stem cell isolation and processing. ZCRM is an Indo–Japanese joint venture claiming to use technical know-how from a Dr Takeshima of a therapy institute in Japan[3] for the treatment of diseases using adult autologous stem cell therapy. But in the second year after the establishment of XIRM, the X Group became self-sufficient by upgrading its stem cell research facilities. After it started conducting stem cell isolation, processing and clinical administration independently, the technical tie-up with ZCRM came to an abrupt end.

The XIRM public relations officer claims that the centre has provided therapy to 470 patients over the last two years for a variety of disease conditions, including spinal cord injury, liver cirrhosis, cardiac infarction and Alzheimer's disease. XIRM receives patients from 20–25 different countries, including Australia, New Zealand, USA, Japan, Pakistan and Sri Lanka. Patients from foreign countries are charged 15–20% more than Indian patients, reportedly because they require extra care during the treatment period. Embedded in the infrastructure of the rapidly expanding medical sector of this city, the X Group's network of hospitals grows applying the 'hub and spokes model', creating fruitful links with other research hubs, conquering new methods, and recruiting new patients from locations near and far. As a result, distinctions pertaining to the private and public, such as public university and private enterprise, and pertaining to the Indian and foreign, such as home-grown and global knowledge, and Indian and foreign healthcare provision, are increasingly blurred.

B. Setting 2, Bangalore: The Y Group of Hospitals

The Y Medical Institute, part of the Y Group of Hospitals in Bangalore, is considered to be a leading centre in stem cell research and therapy in India. Bangalore, capital of the state of Karnataka, has transformed itself over the last decade from 'Garden City' to India's Silicon Valley. Apart from information technology, Bangalore is home to leading biotechnology companies, such as Biocon, and numerous start-ups. It houses some of the world's most advanced R&D centres, such as the Indian Institute of Science, the National Centre for Biological Sciences, the Jawaharlal Nehru Centre for Advanced Scientific Research, the University of Agricultural Sciences, the Central Food Technological Research Institute, and the Institute of Bio-informatics and Biotechnology (Vijay, 2007). The Y Institute of Regenerative Medicine (YIRM) is part of Y University located in Bangalore under the Y Education Group—a leading education provider in India's private sector. YIRM inaugurated itself in August 2007 to become a leading centre in stem cell research and education in India. Under the same Y Group, a commercial entity called Stemline Research Private Limited[4] grew into a leading stem cell company in the field of regenerative medicine. Stemline claims to be the only good medical practice-compliant stem cell research facility in the private sector in India, vetted by the Indian Council of Medical Research. Stemline advertises itself in its policy programme as committed to the delivery of 'bench to bedside' therapy.

In collaboration with approximately 20 corporate hospitals in India, Stemline is conducting experimental stem cell therapies as a 'proof of concept' using bone marrow-derived mesenchymal stem cells in patients suffering from various diseases, such as Parkinson's disease, spinal cord injury, critical limb ischaemia, avascular necrosis, motor neuron disease, end-stage liver disease, psoriasis, vitiligo and myocardial infarction. Stemline also has a branch in Malaysia, which aims to promote stem cell activity locally and focuses on cutting-edge research, therapeutics and therapy in the field of regenerative medicine. A news report published in October 2008 in India announced the launch of therapy facilities at a centre in Kolkata and hospitals in the state of Kerala, and the establishment of tie-ups with 30–40 hospitals in India within 18 months to conduct stem cell therapy. Similar to the X Group, the Y Group ties together private and public research facilities and national and international efforts in its activities to promote stem cell therapies.

Bionetworking and Patient Recruitment

This section shows how demands for scientific quality elsewhere in the world are at the same time exploited and neglected to offer low-quality treatment to hopeful patients from abroad and at home. Stem cell researchers and therapists can take advantage of the fact that in regulated affluent societies great efforts and capital are regarded as essential for establishing conditions for good clinical practice and evidence-based therapies. This has created a niche for the provision of relatively inexpensive therapies that have not been acknowledged internationally. Below we describe the roles of patients, providers and regulation in the recruitment and provision of stem cell therapies before relating their inter-relational dynamics to the broader concept of bionetworking.

Desperate patients and their relatives, especially after having tried conventional therapies, are easy targets for this novel and non-standard therapy. Therapy providers in the settings of the X and Y Groups recruit patients by giving them the impression that stem cell therapy can help with varying disease conditions, and keep known information on the therapy to a minimum. There are various issues at stake here. One is the moral and medical justification of intervention for terminally ill and incurable patients. Another is the issue of providing stem cell therapy to patients whose medical conditions do not warrant it. There are also issues of taking proper informed consent, involving the explanation of medical procedures and risk–benefit analysis (Kiatpongsan and Sipp, 2008; Pandya, 2008).

A. Recruit Urges and Efforts by Telephone

XIRM receives patients from all over India and also from 20–25 other countries. Indian patients choose this hospital on the basis of what they hear through wider media publicity and through referral hospitals. XIRM is part of the X Group, which uses the 'hub and spokes' model to supply patients to its 'hub', or the super-specialised hospitals based in the southern part of Chennai. In general, the patients who come to XIRM have different backgrounds. Domestic patients are drawn from two groups: patients who receive reimbursement for their after-treatment medical bills through their job in the public sector or in large private enterprises, or patients from the upper economic strata of society who can afford to pay for medical expenses out of their own pocket. Patients from abroad are usually referred through large networks that X general hospitals have with clinics and doctors based inside and outside India. Foreign patients can be divided into two groups. One group consists of patients from low- and middle-income countries, such as Pakistan, Sri Lanka and countries in the Middle East, where stem cell therapy and related technological infrastructure are not available; the other group of patients consists of medical tourists, who come from developed countries where the therapies and the associated technologies are present, but are not made available because of stringent regulations.

Even though service providers claim that they provide therapies to poor patients, a look at recruitment procedures, medical expenses and other related costs makes this hard to believe. It is important to note that keeping up the business of stem cell therapy is not easy owing to the considerable investment required for capital and manpower. As a co-ordinator at XIRM puts it,

> It is not easy to get patients who can pay such big amounts, so there is always pressure on us...There is always pressure on us to meet the target number of patients per month. After all, we have to take care of our salary, as well as the investment in state of the art infrastructural cost.

The need for high-end technology-based infrastructure, the novelty of the therapy and the financial situation of the patients induce service providers to develop special recruitment networks. There is no single procedure that these centres follow in recruiting patients. The various social and geographical backgrounds of patients disallow this. Furthermore, it is hard to gain access to patients under treatment, as patient confidentiality provides protection to both patients and provider. Patients are recruited actively, but largely not openly, which makes it challenging to gather trustworthy data on recruitment methods. However, participant observation at one of the centres, discussion with centre officials and communication with patients outside the centres gave some insights into how patients are recruited. During a third visit to the co-ordinating office of XIRM's stem cell department, a telephone call came for the co-ordinator, Mr S.R., during our conversation. The following is the co-ordinator's side of the conversation:

> From where are you calling? Oh! Kerala...
> (...)
> No, we are not providing the therapy for kidney-related problems right now. But, research is going on, and very soon the results will be out and...
> (...)
> Yah! Yah! You can just keep on track; just keep on track in newspapers.
> (...)
> Yes, yes, usually the results will be declared through newspapers.
> (...)
> Yes, you can contact me after...say 15 days.
> (...)
> Yes, just listen. I am almost sure that the on-going study will be positive and we will be able to provide the therapy. Yes, the research is going on at our hospital. We have advanced research facilities and we are also watching the results of studies at other places...But, I cannot say for certain. Yah! Yah! Please call me after two weeks.
> (...)

Yes, you have to come with all your past medical reports and our experts will examine you and then decide if you can take the therapy or not. No, no. It is not painful. We have some experts who will administer it. Don't worry about that.

(...)

See that aspect we cannot tell right now, it depends how much you need, the quality of cells and all that. But, since you will be the first case, we will charge very nominal. But, please call me after two weeks.

(...)

OK. Thank you. Hear [from] you after two weeks. Bye.

B. Recruitment in Person

On the same day, another incident took place, which yielded more knowledge about this stem cell centre's recruitment procedure. A man arrived searching for the stem cell co-ordinator at XIRM, Mr S.R., He looked worried and tired. The man was from Rajasthan, some 1500 kilometres from this centre, and he wanted to know more about stem cell therapy for his younger brother who had suffered from a spinal cord injury for the last three years.

The discussion between Mr S.R. and the man was as follows:

The man: Hello. I am...I am from Rajasthan and I want to know about what kind of therapy you are providing at this centre.

Mr S.R.: I see. We provide stem cell therapy for many untreatable diseases, but...What is your problem?

The Man: No, my younger brother met with an accident some three/four years back and is suffering from spinal cord injury.

Mr S.R.: Yes, we do provide therapy for spinal cord injury, and I tell you, most of our patients are here for spinal cord injury. We have expertise on that. Since you said your brother's accident history is only three years, I think it will be very receptive to the therapy. It is an ideal case for intervention. I tell you...

The Man: He has injury at [the] C-5 and C-6 positions of his spine. We have already been to many places, including Dr G's clinic in Delhi, where he has taken embryonic stem cells for six months, but there is not much difference. We have been to a doctor in Ahmadabad and one in Jaipur. I have also talked to people at Y Hospital in Bangalore...

Mr S.R.: What did they say?

The Man: They also said their therapy will be receptive to my brother's condition, but they are charging a lot of money.

Mr S.R.:	How much they are charging?
The Man:	Yah…It's a lot, we cannot give that much, and we have already spent a lot at Dr G's clinic in Delhi…
Mr S.R.:	See, we charge the minimum in the market. Our results are the best. Of course, if you want, you can talk to our medical team.
The Man:	But, I want to know, how do you insert the stem cells. Is it at the injury site, or, I heard something like opening the spine…
Mr S.R.:	Yah…see, it all depends on your brother's condition. Our experts will decide whether to inject it at the injury site or to open the spine and inject. But, we [use] the least invasive in our techniques and methods; see, we are the best.
The Man:	What will be the price?
Mr S.R.:	See, first you decide and then we talk about that. That is not the big issue. The thing is that you will get the best treatment. We charge something around 2.5 lakhs [around $5500]. But, I will talk to the management, and will make some concession for you, since you have already spent a lot of money at other centres. It will be on humanitarian ground[s]. Another good thing about our centre is that we just give the stem cells in one or two doses, or maximum three doses, not for so many times as the other centres are doing. They are just looting people. We are very clear about that. If it works, it will work in one or two times, or else it does not work.
The Man:	What is the success rate?
Mr S.R.:	See, it varies from condition to condition, since your brother's injury is relatively fresh, he will have better success rate. This is what I think. But, you know, it all depends on God! But, at our centre we have [a] nearly 75% success rate for many patients.

These conversations between relatives of patients and the representative of the service provider give us an idea about the sense of urgency on both sides. Urged to recruit patients, the co-ordinator wittingly or unwittingly takes advantage of the dilemma in which patients and their relatives find themselves, placing their money on hope offered and a chance of survival. Speculation on the health, schooling and financial capacity of the patient connect the biological needs of patients and the financial sustenance of co-ordinators in, what has become, the business of bionetworking.

C. *Recruiter Patients*

Another form of attracting patients is by using what we call 'recruiter patients'. The role of recruiter patients is more prominent in the private sector and among individual practitioners in comparison to the public sector. Patient recruitment requires a well-defined strategy that varies across service providing centres and hospitals depending on their size, infrastructure and networking tactics. The use of patients to attract new clients is an integral part of the private service provider's scheme. A public relations officer of a multi-specialty private hospital in Chennai, which claims to have provided stem cell-based therapy for over 400 patients in the last two years for various medical conditions, explains the rationale of using recruiter patients:

> People will come to you only when they see the results for themselves and they are convinced about it. They want to see it before they believe it. Or at least they want to listen first hand. You know, they have to spend so much money. Therefore, they should have a chance to see for themselves and listen to the experiences of patients who have benefited from the treatment. For that reason we arrange meetings with patients or provide their contact details so that they can directly interact. For us, the patients who have received good results from the treatment are our best advocates. You can say they are our ambassadors.

The views of patients and their relatives about the role of recruiter patients varies, usually depending on the therapy provider they are associated with. Some patients and relatives believe that recruiter patients have played an important role in their decision to go for stem cell therapy. They feel that the first-hand narrative of the recruiter patient helped them to make an informed decision about the risks and benefits involved, about the quality of service and about what to expect from the treatment. One patient, who had recently received several doses of embryonic stem cell therapy at a private clinic based in New Delhi, said the following:

> We came to know about this clinic in Delhi from a patient who had received stem cell-based therapy here for spinal cord injury. Upon our request, the clinic provided his contact address to us. The clinic even arranged a meeting between the cured patient and us [family members]. After seeing the patient's improvements, his self-confidence and after listening to his narratives we decided to receive the therapy. His narratives and experiences were very convincing (transl. by *Patra*).

Another patient had a different experience. Mr V believed that the contact addresses and names of patients provided by the clinic/hospital for pre-service interactions were fake, and that they constituted manipulative acts by the clinics concerned. He claimed that recruiter patients work as agents for private providers and that their main job is to attract as many patients as possible into the ambit of therapeutics:

> I urge people not to fall prey to the trap that these hospitals are adopting. To my utter shock, the doctor and the whole LL hospital system are into fraudulent activities. They use 'old patients' and their false testimonials to influence clients. When I asked for contact numbers of some patients, I was given two contact numbers by the public relations officer of the stem cell department at the hospital. The two patients I contacted over the phone had a very high opinion about the treatment results and hospital facilities. But they showed one or the other pretext to avoid a personal meeting with me. Later on I found out that the numbers provided belong to the staff of the same hospital and the patients were not real but fake. I felt cheated (transl. by Patra).

Private hospitals involved in stem cell therapy services try to maximise patient intake by the strategic use of recruiter patients: their symbolic value as cured patients is used to entice patients to undergo unverified therapies. Two examples show how when the health symbolism of recruiter patients is questioned, the meaning of healthcare provision in general becomes problematic. The first example of Mr Patel is based on a web-based narrative by a patient who had received adult stem cell therapy for his spinal cord injury, and the second case is based on Sleeboom-Faulkner's meeting with the same hospital personnel. The cases illustrate how stem cell therapy service-providing centres in the private sector use the symbolism of cured patients and forge the role of recruiter patients strategically as part of their bionetwork.

Mr Patel, a spinal cord injury patient from Gujarat, contacted XIRM Hospital for possible stem cell therapy after he had been unsuccessfully treated for the ailment elsewhere. He came to know about the hospital from a newspaper article and subsequently from the hospital's website. Desperately, he wrote an email to the head of the stem cell section at XIRM Hospital, Dr R, for guidance. Dr R advised him to speak to two patients who had been successfully treated for a similar injury at XIRM Hospital. Upon contact with both patients, Mr Patel was given positive feedback about the treatment and the hospital. He was influenced by the feedback and decided to have the therapy at XIRM Hospital. However, after a while, he became suspicious of their communication, including that of the co-ordinator of the hospitals, who managed the treatments, and one of the patients with whom he had interacted earlier. Eventually, he became convinced that the voices of the co-ordinator and the patients were one and the same, and that he had been scammed. On his Internet blog the patient gave the following statement about the incident:

> A while back I spoke to Mr. SR who is a coordinator of the stem cell department in [XIRM]. Something was very fishy; the voices of Mr. SV and Mr. SR seemed very similar. So I went to the website http://www.stemcell-india.com/ and checked the Contact us section http://www.stemcell-india.com/contact.htm. I was so hurt to see the Mr. SR's number on this page and Mr. SV's number on the above email were one and the same (taken from the web-blog of Mr Birju Patel, dated 8 February 2007).

Not surprisingly, the magic attraction of the recruiter patient disappeared in the mind of the disillusioned patient.

When visiting XIRM Hospital in Chennai, Patra met with the co-ordinator and physician-in-charge of the stem cell therapy department. Upon requesting to meet patients who have had successful stem cell transplantation, the contact numbers of two 'patients' were provided. No response was obtained from one of the numbers and the other was found to be a recruiter patient who gave very positive views about the treatment and the hospital's medical services. However, the patient did not want to meet personally, as he lived far away from Chennai. After cross-checking the contact number, it was found to be identical to the number of a member of the stem cell department of XIRM Hospital. The identity of the patient had been forged and the entire episode appeared to be stage-managed. The two examples show a similar story in which the authenticity of recruiter patients is disputed by service-seekers when they discover that some service providers provide details of people who pretend to be and act as cured patients. When the pretence of the service provider is revealed, stem cell therapy in itself becomes disputable, which has consequences for the acknowledgement of the integrity of stem cell research in general.

With the growing public expectations of stem cell therapy and the widening access to global information flows, stem cell therapy-providing centres have come to play a crucial role in clinical stem cell applications. The proliferation of unverified stem cell-based medical treatments has complicated the decision-making of patients and their relatives with the increased use of recruiter patients. Next to recruitment by telephone and in person at the clinic, recruiter patients have become an additional means through which bionetworking activities are executed by service providers: they exploit relations of dependency and financial and health uncertainties around the globe.

Case 17: Bionetworking Between Japan and India[5]
(*with Prasanna Kumar Patra*)

Stem cell research has been a focal point of the life sciences in Japan's Millennium Project, which aimed to boost economic development and therapies for its rapidly ageing population (Sleeboom-Faulkner, 2008b). Government investments in stem cell research have been substantial, expressed in the establishment of large stem cell hubs at RIKEN institutes (Tsukuba and Kobe), Keio University and Kyoto University. In the midst of this upbeat policy-making context of innovation, investment and high expectations, uncertainty reigns over the extent to which research funding will lead to results and experiments can be translated into effective stem cell therapies. Great pressure on scientists to produce international science publications and to enter collaboration with industry to develop effective therapies (Saegusa, 1999) has led to the concern that scientists may be tempted to get involved with dubious forms of human experimentation in both India and Japan. At issue in this case study are the questionability of the scientific basis, the medical efficacy and the regulatory permissibility for using adult autologous stem cell therapies, which make use of the patient's own stem cells, as well as the ambiguity with which adult stem cell therapy is promoted as embryonic stem cell therapy in many hospitals and clinics. Although adult autologous stem cell therapy is widely considered to be free from ethical problems and thought to have great therapeutic potential (Khan, 2007), the international scientific literature recommends embryonic stem cell therapy only when used in a strictly monitored environment (Partridge, 2003; Strauer and Kornowski, 2003; Wainwright et al., 2006; Passier et al., 2008). Similarly, the International Society for Stem Cell Research makes a clear distinction between routinely used and experimental stem cell therapies. Bone marrow stem cell therapy is a well-established treatment for blood cancers and other blood disorders. But new clinical applications for stem cells are currently being tested for the treatment of musculoskeletal abnormalities, cardiac disease, liver disease, autoimmune and metabolic disorders and other advanced disorders. These new therapies have been offered only to a very limited number of patients as a last resort and free of charge. The uncertain nature of these stem cell therapies led the Indian Council of Medical Research (ICMR) to develop research guidelines to guide the development of new stem cell therapy testing (DBT-ICMR, 2007). But, as the guidelines are 'soft' and not clearly formulated, enterprises regularly ignore them (Salter, 2008; Bharadwaj and Glasner, 2009; Patra and Sleeboom-Faulkner, 2009b).

Many clinics in India practice on and sell unvetted stem cell therapies to patients who cannot find alternative treatment methods. The scale of this research is such that the ICMR does not have the capacity to control violations (interview, ex-official ICMR, N. Agarwal, July 2010). At the same time, belief in the progressive image of modern science raises hopes that crucial steps will occur in research fields and therapeutic areas where foreign scientists fear to tread. The advanced aura of Japanese technology, then, is mobilised to attract patients in India. In Japan, problems of a different nature arise. The development of interim guidelines for somatic stem cell research in a clinical setting were only finalised in July 2006, despite years of preparation and debate among scientists, patient groups and disability groups (MoHWL, 2006). Guidelines for stem cell research in Japan are also loose, but social control through gate-keeping bureaucrats and self-regulation is strict; researchers experience the guidelines as limiting because they require researchers and hospitals to go through lengthy processes of application for permission (Slingby et al., 2004).

Japanese stem cell researchers are aware of the problems associated with premature clinical applications. Time and again, interviewed stem cell scientists insist that it is not yet clear how stem cells regenerate and repair tissue, how adult stem cells work in different parts of the body, how to standardise the cells and how to improve the efficacy of the therapy. In one instance, the company J-Tec, which tried to regenerate skin using autologous bone marrow stem cells, had to wait for more than seven years before its trials were recognised (interview J-Tec, 9 November 2008). Other researchers complained about not receiving permission to use mesenchymal stem cells in the mouth (interview Ushi, 22 October 2008). Although these complaints have prompted the Pharmaceuticals and Medical Devices Agency (PMDA) to reorganise application procedures, and the Ministry of Education, Culture, Sports, and Science and Technology to reorganise science research around thematic and geographic regions, bureaucratic challenges are far from resolved (interview, Hara, 7 November 2008).

The challenges in conducting human subject experiments for new therapies in Japan do not just relate to social control, the media and bureaucracy, they also relate to patient recruitment. Japan, as an affluent welfare society, provides a high level of healthcare for the vast majority of its population, and supports adult stem cell therapies in the case of untreatable and life-threatening diseases. But only when no other drugs or therapies are available do Japanese patients seek refuge in experimental medical trials for which they are not covered by insurance, or travel abroad to look for available/cheap medical assistance (Yuasa, 2007; Fukuda and Nakai, 2008). In less wealthy countries that lack universal healthcare coverage, such as India, experimental clinical research may seem the best treatment option for many patients (see Patra and Sleeboom-Faulkner, 2011). Researchers who are waiting for approval from regulatory organs in Japan can be tempted to conduct research in countries that have a large supply of patients at the bottom of a biohierarchy (interview, Dr Egawa, 17 November 2008). Regulatory and socio-economic discrepancies, then, have opened up a new transnational space for clinical research and have become the playing field for bionetworking strategies (Patra and Sleeboom-Faulkner, 2009a).

Dr Ghandi's Transnational Bionetwork

Dr Ghandi is the Indian founder of ZCRM, an Indo–Japanese joint venture institute based in Chennai. Ghandi is a research manager, scientist and businessman all in one, and knows how to combine his various functions across borders. Ghandi previously worked through Z Bio Sciences Private Limited, established in May 2000 with Japanese equity. He engages in diverse fields of activity: healthcare; R&D in the life sciences and plant biotechnology; herbal medicine; technology transfer between Indian and Japanese institutes; and importing and marketing of surgical devices, biotechnology labware and products for agriculture. Through his work in Chennai, Ghandi expanded his links with Japanese researchers, scientists, medical doctors and biotechnology companies interested in investing in stem cell therapy in different areas: clinical trials, herbal medicine, and the selling of medical equipment and chemicals. His career follows life science trends: while in a position at Yamanashi University for over a decade, Ghandi became the lead researcher and director of the Chennai laboratory of ZCRM and director of the ZCRM in Tokyo (Z International), which was registered in November 2008. ZCRM started as Standard Private Company Limited in 2000 with Japanese equity, after which it became Nichi-In Drug and Device Private Limited. Later, it became Nichi-In Bio Private Limited and, finally, in 2005, it became ZCRM.

Over time, Ghandi has signed many collaborative agreements between research institutes in Japan and India. Those signed with Japanese institutes pertain to science research, while those with Indian institutes pertain mainly to translational research and therapeutic applications.[6] Thus, ZCRM has few (basic) science collaborations with Indian institutes (only the Department of Biotechnology of Acharya Nagarjuna University is mentioned). In Japan, however, Ghandi has research collaborations with Waseda University (Department of Polymer Sciences, Tokyo), Mebiol Inc. (Tokyo), the Kawamura Institute of Chemical Research (Yamanashi University Faculty of Medicine), Yamaguchi University Faculty of Medicine (Ubu) and the Biotherapy Institute (Tokyo). Ghandi's network, then, displays a rough division of labour between scientific research in Japan and translational research in India. Ghandi opened ZCRM's Tokyo office, headed by Mr Yamazaki, on 7 January 2008. The office serves to promote medical tourism in India by tapping into the Japanese market, developing collaborations with Japanese institutes and industry, and dealing with intellectual property-related issues in Indo–Japanese research. Ghandi claims that ZCRM's research on stem cells in fields such as ophthalmology, hepatology, orthopaedics, dermatology, dentistry and oncology takes advantage of advanced Japanese technologies. It also boasts, to its credit, several firsts in clinical studies using stem cells for spinal cord injury and liver cirrhosis. Furthermore, Ghandi claims that ZCRM is the first institute in India to use autologous immune enhancement therapy for cancer, in which a patient's own natural killer cells and T lymphocytes are used to curb cancer cells (ZCRM's website, accessed 22 January 2009).

Ghandi provides stem cell isolation/expansion services to Indian hospitals. Formally, these hospitals work according to the guidelines of the Institutional Ethics Committee and Institutional Committee for Stem Cell Research and Therapy (IEC/ IC-SCRT) stipulated by the ICMR. The guidelines permit the use of autologous stem cells in research and treatment, provided the protocols are approved by the IEC and IC-SCRT, and registered with the National Accreditation Committee for Stem Cell Research and Therapy (DBT-ICMR, 2007). But, as there is no adequate supervision over IEC/IC-SCRT, and, as pointed out by Ghandi, the National Accreditation Committee has not yet been formed, and regulation does not stand in the way of research and therapeutic applications (telephone interview, Ghandi, 17 January 2009). To avoid becoming entangled directly in the ethical issue of applying unauthorised therapies, Ghandi uses his collaborative networks to recruit and maintain a pool of patients. ZCRM recruits, treats and charges patients indirectly. According to the co-ordinator of ZCRM:

> The concerned collaborative hospitals recruit patients, treat them and charge them for the services. We do not directly charge the

patients. We charge the hospitals for our services. How much we charge, I cannot tell you. Because, you know, as per the ICMR guidelines, it is an experimental therapy and for experimental therapy we should not really charge...We charge for the laboratory expenses. It is very nominal (interview, 5 November 2008).

Ghandi avoids applying for clinical research authorisation and directly charging patients through bionetworking, illustrating his biomedical ethic. Ghandi did not take the time to meet Patra in India. Instead, he told her to meet the public relations officer, Mr Kamal, at the down-market office and laboratory at Vijaya Hospital compound in Chennai. This makeshift clinical hut operates in the not-quite-illegal healthcare interstices where control is either absent or ineffective. Mr Kamal was quiet and taciturn, and only conceded to share a PowerPoint presentation of ZCRM. Any questions were directly referred to the director of ZCRM. ZCRM's activities in India, according to Mr Kamal, cover diverse areas: first, it performs R&D at lower costs than its counterparts in Japan and applies for Indian patents for basically unaltered, but now tested, technologies; second, it stores various types of stem cells, and aims, with external funding, to establish the world's first corneal endothelial stem cell bank (CESBANK); third, it finds users for Japanese technologies, whose fees they reinvest in the enterprise; fourth, it showcases selected patients to advertise the effectiveness of its experimental therapies, to attract patients and new collaborations for stem cell tourism, both in India and in other parts of Asia, such as Malaysia, Singapore, Indonesia and Brunei, where the Z-Asia Centre for Regenerative Medicine has been set up; fifth, it provides international training programmes via the web to contacts in Canada (with the University of Toronto, McMaster University and the University of Ottawa) and Asia; and, sixth, it advertises itself as having 'excellent networking at all levels (local, national and global) in all three areas (basic, translational and clinical studies)'. According to the public relations officer of ZCRM in Chennai, ZCRM, with various collaborating centres, has treated more than 150 spinal cord injury cases (in six hospitals), 25 cardiac disease cases (two hospitals), numerous peripheral vascular disease cases (two hospitals), 35 liver cirrhosis cases (two hospitals) and two oral sub-mucosal fibrosis cases (two hospitals).

ZCRM competes with other research centres, especially the Manipal Institute for Regenerative Medicine in Bangalore. Among its major achievements, the officer claims that ZCRM has been the first in India to perform immunotherapy for cancer patients; to have performed stem cell therapy for diabetic ischaemia patients; to have performed stem cell therapy for spinal cord injury; to have two publications in stem cell therapy and ten clinical presentations; and to have two patents (applied for) on stem cell and regenerative medicine for the isolation of stem cells. But the only publication evidence of work related to this list of claims we found was based on conference proceedings in the *Journal of Stem Cells and Regenerative Life Sciences and Biotechnology*. This is a non-peer-reviewed, free online journal, with its editorial office and headquarters in Chennai and Kofu respectively—the Indian and Japanese homes of Ghandi. In short, Ghandi cultivates and maintains a large network in Japan and India that provides 'Japanese' stem cell technology to Indian therapists, whom ZCRM first trains and then charges for stem cell applications to patients, thus avoiding ethical entanglements, while closely following the progress made with treatment methods and technologies.

Japanese Scientists

Ghandi taps Japanese resources in various ways. On the one hand, he uses newly developed technologies that are not yet authorised by the government to license to hospitals in India. On the other hand, he capitalises on the need of Japanese research centres to experiment with their technologies on patients. The shortage of patients for clinical experiments in Japan derives from the high standard of treatment covered by National Health Insurance, which makes it less attractive for patients to pay for new experimental treatments in private. Thus, in 2008, one government-sponsored research centre in the Kansai area was waiting for a general hospital to be built adjacent to it in order to close the gap between newly generated technologies and clinical studies. In the same centre a meeting took place with a high-tech institute in India. According to one of the researchers present, a leading researcher suggested that

> There is a good opportunity for collaboration, because the possibility of doing clinical trials in Japan is usually limited by the number of patients. The doctor there answered that 'We have a lot of patients, and we have actually young patients as well'. That would be very interesting for Japanese studies. The limitation there is probably money and other aspects. He said, 'We have enough money here, but not enough patients'. But I do not know if anything developed out of this (interview, Dr Panto, 1 December 2008).

Although interviewees admit that they commonly entertain the idea of bartering between patients and technology and investment (e.g. Egawa, interview 17 November 2008), such exchanges are usually referred to as 'collaboration' and 'scientific exchanges', and sometimes as 'providing help'. Ghandi speaks of such exchanges in terms of charitable activities, providing cheap treatment to Indian patients, and aiding Japanese patients who cannot find treatment in their own country, in both cases through reliable Japanese technologies in India (interview 5 November 2008; ZCRM's website, accessed on 22 January 2009).

Ghandi is especially aware that testing the high-profile induced pluripotent stem cell (iPS)[7] technology developed by Yamanaka Shinya from Kyoto University will require numerous patients. At the same time, it would be a great asset to young Indian researchers. For this reason, Ghandi has arranged to have researchers from India trained at Kyoto University, and has established connections with leading researchers working on future clinical applications:

> I have met Dr Yamamoto and Professor Takeuchi a couple of times. When we send some researchers there, they will be training them. Now he has proposed that we should start collaboration, send someone from India to do iPS research, and doing iPS research in India...Now we have a meeting in Tokyo with him on 5th or 6th of March to make an agreement...He is not sure if Kyoto will send its people to India, but Yamamoto has already been to India for a conference for a few days, so he already knows what is happening there. He was really impressed by the young powerful youth in India (interview, 5 November 2008).

At a meeting with Professor Takeuchi in Kyoto, however, Professor Takeuchi flatly denied ever having heard of ZCRM or of Ghandi. When asked what he thought of conducting experimental research abroad, Professor Takeuchi responded:

> It is wrong. Using Japanese people's tax to create new applications not using it for Japanese people, but selling it to Americans...No matter how you look at it, it is abnormal. Not keeping the diseases of the Japanese people in mind when developing new applications is strange [*okasshi*] (interview, 30 October 2008; transl. by Sleeboom-Faulkner).

This response was an attempt to steer the conversation away from Ghandi while trying to express commitment to Japanese science. Though Professor Takeuchi showed his Japanese patriotism vis-à-vis the USA, he had other concerns when considering testing 'Japanese' technologies in India:

> If you can do good research and use proper equipment for the research in India, and if you can use it in Japan as well, then it would

be OK. But [as they are different] you would have to do the experiments once again for the Japanese (ibid).

He did not deny the possible acceptability and even advantage of having a link with Indian patients. Experimental data obtained in India are useful because they can provide arguments for permission from the PMDA to test new treatments in Japan. Even though the International Conference on Harmonization of Technical Requirements for Registration of Pharmaceuticals for Human Use has reduced the need to repeat research through agreements that recognise the validity of test results in member countries, Japan has been able to negotiate that, to a certain extent, clinical trials and experimental research conducted abroad have to be repeated in Japan for disputable reasons related to race and environment (Kuo, 2008). Consequently, and as argued by Japanese scientists such as Professor Takeuchi, the availability of data from abroad can count as favourable evidence for seeking permission to conduct less extensive clinical research. Professor Takeuchi's refusal to admit any knowledge of Ghandi, however, is a sign that he is unsure about whether Ghandi will bring scandal, become a threat to scientists of good repute or facilitate research. Similarly, a senior researcher in the same institute working under Professor Takeuchi, Dr Yamamoto, who actually had participated in a conference organised by Ghandi, denied any connection with Ghandi apart from knowing him as the organiser of a scientific conference held in India. Although Dr Yamamoto kindly explained to me all about his work on mesenchymal stem cell research applications to knees (interview, 27 October 2008), any question remotely related to India was diverted to stem cell research applications of iPS. Little did he know that Ghandi would place a photograph of him taken in India on the front page of ZCRM's bulletin.

Confronting a scientist with the application of their technology to cases in India led to other ambiguous reactions. Ghandi kindly introduced one of the authors (Sleeboom-Faulkner) to Dr Umemura, whose immunology therapy, called artificial immunology (AI), is licensed to Indian clinics. Dr Umemura was asked questions about the following description of the case of Mr Ali, which appeared on the XIRM website:

> The 25-year-old Mr Ali, who sustained an injury to his spinal cord in October 2006 at a construction site in Abu Dhabi, became bound to a wheelchair. In December 2006, he was admitted to the XIRM (of the SL Hospital Group), Chennai, for reparative surgery. The hospital had just signed a MoU with ZCRM, and it was decided to

use the autologous stem cell technology developed by Dr Umemura of AI-Therapy Institute, Tokyo, Japan. Stem cells were isolated from his iliac bone, expanded and injected into the space around the injured site of the spinal cord...Two months later, Mr Ali was able to withhold his urine for up to 2 hours and could walk independently. At a press meeting in February 2007, graced with the presence of Mr Otsuji, Consul of Japan, Dr Raj presented the case to the audience.

When asked about the method he used on the spectacular case of Mr Ali's spinal cord injury, Dr Umemura said:

Even though the technology is based on publicised science, this [applying his method to spinal cord injury] seems to be out of control. I created the method for cancer application. Dr Ghandi, as far as I could see, was quite different. I cannot support the use of these technologies in India, for I do not know if accidents will occur. But if it goes out of control [causes a scandal], also in Ghandi's case, I withdraw my support (interview, 18 November 2008; transl. by Sleeboom-Faulkner).

Although it is not clear if Dr Umemura benefits from the licensing fees for the use of his technology in India, any scandal around his technology would damage his reputation and partnership with other Japanese clinics. When asked, Dr Umemura clarified that his technology was not recommended for SCI cases. Furthermore, the AI technology used by ZCRM is, for financial reasons, limited to using so-called natural killer cells, while that in Japan is combined with cytotoxic T lymphocyte cells.[8] Moreover, the therapy, even if provided in its entirety, is not acknowledged as standard treatment in Japan.[9]

A meeting with Mr Ali could shed light on the effect of the therapy. But when SL Hospital was asked for Mr Ali's contact information, it turned out to be 'lost'. Providing an inferior therapy in India would go against the spirit of Japanese regulation, which does not condone the provision of therapies abroad if they are not approved for Japanese people (interview, Egawa, 17 November). Nevertheless, Dr Umemura almost certainly knows about the use of such applications in India, as photographs on the Internet attest to his visiting the clinic where the therapy was applied to spinal cord injury patients—Ghandi needed the Japanese scientists as the faces of the technology he licenses out to Indian therapists for application to patients in India. Ghandi's task is to emphasise the charitable aspects of their collaborations, to celebrate international relations, and to declare the cause of science and developmental aid, while underplaying the licensing fees, experimental testing and risk to patients.

In the case of ZCRM we observed, again, different forms of patient recruitment. The faces and names of scientists and technologies from countries known for their 'advanced' life sciences and healthcare are deployed to attract patients, who either are treated for free or pay large sums of money for treatment. The bionetworking skills of science managers include the use of background knowledge of patients to determine price, therapy and risk assessment. Bionetworking activities here constituted a leading science manager with access to contacts, hospitals, investors and personnel, in both Japan and India, and the relevant knowledge about differences between Japanese and Indian cultures, regulation and healthcare provisions to succeed in recruiting patients, applying for licenses, establishing connections with authorities and regulators, such as the IMCR, Japanese business license authorities and the Japanese embassy in India. It was not just the awareness of material differences between rich and poor, and the awareness of where healthcare and technology are available; it is, in particular, the emphatic ability to weigh the risks and opportunities of different patients against the risks and opportunities of bionetworking activities. This is done not just by hyping the promising nature of new technologies, but by playing on the particular needs particular patients have considering their family background, financial standing, access to knowledge, medical prospects, and perception of science and matching those with what can be offered in terms of treatment.

Conclusion

Science managers involved in recruitment by telephone, in person and through recruiter patients link together service providers, patient groups and their families, and stem cell therapy technologies across local and global spheres of practice. The science managers observed in the case studies in this chapter, in terms of bionetworking, require the ability, first, to tap into the structural inequalities of a socio-economic nature; second, the hyping of biotechnology applications catered to the needs of certain patients; and, third, the differences that have grown historically, including those in regulation, healthcare and socio-cultural traditions. Thus, the ability to purchase, to access healthcare and to understand medical terminology are important clues to a bionetworker's decision about which therapy to offer and how to advertise it.

Hyping of the therapies takes place through advertisements on the Internet and media, and by acquiring the approval of persons or institutions with authority. Whereas the hyping of new biotechnologies facilitates the belief in new therapies by managers and scientists that are confident in their ability to help cure patients, there are other ways through which therapies reach the market. As described above, there are cases in which recruitment methods are fraudulent. This is the case when the strategic mobilisation of recruiter patients as human bait aims to attract patients to controversial experimental stem cell therapies. Furthermore, the institutional connections and traditions in which biotech therapy providers operate may financially condition the ability to provide therapies that conform to official guidelines, or the rules of scientific research and good clinical practice. Under pressure, such conditions may form a temptation to bend guidelines and interpret them creatively. The institutional organisation of science and healthcare, such as was the case in the 'hub and spoke' model of the X and Y groups, may lend itself to cover up such creative interpretations, which seem to be imminently reasonable and ethical when patients demand treatment.

There can be a fundamental difference in the intention, planning, procedures and long-term goals that partners in a science collaboration have in mind when they hope to share resources, knowledge and ideas to achieve a scientific purpose, but the dividing line between science collaboration and bionetworking may be impossible to draw. 'Regular' approaches of science collaboration can also aim to acquire profit, fame and IPR, while such collaborative enterprises may not be free from corruption and mismanagement either.

Nevertheless, the concept of 'bionetworking' serves to draw attention to the non-scientific aspects of collaboration. In some forms of bionetworking, as we have seen above, the knowledge of matters peripheral to scientific pursuits are prioritised. Although managerial and scientific activities of contemporary principal investigators of science projects have become increasingly inflated, in principle scientific discovery and the efficacy of biomedical solutions are central to the pursuits of scientists. But only by empirical research into the connections made and steps taken by scientific projects can we acquire an idea of the relative weight put on moneymaking on the basis of non-scientific activities. Illicit bionetworking activities, such as selling therapies under false pretences, may be unethical, but they are not necessarily fraudulent. More insight into the dynamic interplay between biomedical innovation and scientific advancement and its funding is necessary to understand the conditions and driving forces behind bionetworking. Understanding the cross-cutting linkages between life assemblages is the first step.

8
Reframing the Global Moral Economy: Life Assemblages and Research Objects

Introduction

Bioethical ideas, guidelines and institutions are seen as the result of values negotiated in a global moral economy meant to facilitate scientific development and biomedical innovation; the belief in this arrangement forms an important rationale for investment in the life sciences. The life assemblages studied in this book, however, show that a 'shared' global moral economy of the life sciences does not currently exist. This absence of shared moral purpose in the life sciences is shown in the way life science projects and ventures compete and collaborate with different aims. While some aim to find cures for old-age diseases for the population that supports the research, others strive for market share enlargement or fame. The rationale for conducting particular kinds of life science research may resonate better with some research environments than with others.

Attempts to standardise bioethics institutions globally seem to be misplaced. Standardising it cannot yield the same benefits to scientists and the populations they work with in different locations. This chapter aims to deal with the question of how to deal with the diverging problems occurring at the conjunction of the life science and society in different life assemblages. Revisiting terms introduced in the previous chapters—'science collaboration', 'global risk signatures' and 'bionetworking'—I will link the notion of life assemblages to philosopher Gerard de Vries' work on 'research objects'. I will maintain that there is a need to examine life assemblages to tailor bioethics institutions to research objects. The notion of research object incorporates the Aristotelian notion of a 'common good' shared by the political constituency it belongs to. An integration of the various aspects of life assemblages, discussed in previous chapters, will serve to explain how the aim of research in various ways can relate to such a 'research object', which here refers to the research or therapy practice delivered in a community as a common good. In short, I argue that rather than aiming for a global moral economy of biotechnology with standardised bioethics institutions, a better understanding of life assemblages in Asia would equip bioethics institutions to serve research objects more effectively in alignment with locally developed notions of a common good.

Global Moral Economies: Reframing Life Science Ethics in Asian Contexts

In this section I reframe the idea of a global moral economies in terms of the notions developed in earlier chapters and in the context of life assemblages in Asia. This book has aimed to show that moral economies in Asia take on a multitude of shapes and forms, making it both impossible and unhelpful to generalise about the global or Asian moral economies. The notion of life assemblage, discussed in terms of biopower, agency, translations and embedding of bioethics, global risk signatures, bionetworking and frameworks of choice conceptually links developments in the life sciences with regionally disparate values of morality, social organisation and livelihood.

Frameworks of Choice

Rather than describing the global presences of the life sciences as a functionally inter-related web at the service of global health, the term of life assemblage emphasises and clarifies the contested and diffuse nature of 'global moral economies' as both lubricants of international life science transactions and as source of conflict. Although life assemblages share discourses, questions and problems, they also harbour and produce conflicting values, contested activities and disagreements. Such contestation is expressed through various levels of governance and units of decision-making, that is, national government, province, town and family household. State influence may be strong or weak in the formulation and promulgation of bioethical guidelines, but even when authorities support and implement internationally dominant bioethical guidelines, spaces open up where manipulation, reinterpretation and creative negotiation of bioethics and public policies take place to various extents. In this complex, dynamic environment the interaction of social actors, things and modes of governance form the 'frameworks of choice' of a life assemblage. The concept of 'frameworks of choice' attracts awareness to the margins of choice in the changing socio-economic and cultural contexts in which actors operate within or across national borders. It is through these frameworks of choice that actors, including scientists, patients and family households, perceive formal regulation of bioethics and take part in the changing biotechnological practices of life assemblages.

Life Assemblages and Bioethics

Life assemblages connect elements of life science and society that reflect different and broader ethical concerns compared with those salient in the political discourses around research ethics in the life sciences. Thus, it is possible that life assemblages contain practices that seem to be in conflict with the ideas advocated by formal bioethics institutions. As life sciences are 'co-produced' with society (Jasanoff, 2005), formal bioethics institutions form just a fraction of the moral economies underlying life assemblages in practice. Chapters 3–5 showed how bioethical practices in predictive and genetic testing technologies in practice affect ethical issues among the population in a wide range of fields. For genetic testing technologies generate new forms of 'social justice', as they serve some socio-economic groups better than others, and differently affect the availability of choices and 'health needs' of populations—in this case, China, Japan and India—whose life assemblages vary. Varying health needs and health provisions may be crucial to the justification of collective investments into life science ventures. Bioethical guidelines that are in place may be more efficacious when reflecting such justifications.

Moral Boundary-making

An important connective element linking life science and life values into life assemblages is the agency of scientists. Scientists' views of what kind of science is justifiable in terms of the community they are part of and the kind of science they produce is closely interlinked. In Chapters 6 and 7 we saw that interviewed stem cell scientists in China, South Korea and Japan are active users and producers of moral values, and can usually explain the socio-economic importance and aims of their research in both scientific and ethical terms. Scientists in some countries are aware of the differences between the moral justification of guidelines regulating their research and those valid and practised in other countries. Chinese scientists are actively engaged in national moral boundary-making (see Gieryn, 1983, 1999). Drawing moral boundaries facilitates the communication of messages about the scientific values and ethical standards scientists would like to be known for and helps them to negotiate collaborative possibilities within what is perceived as the global moral economy. The existence of national bioethics institutions allows for the transgression of national moral boundaries through transnational collaboration. It enables some cross-border collaborations to select the most favourable conditions that regions, countries or institutes of collaborative partners have to offer, for example funding and knowledge in one country, and loose regulation and affordable expertise in another can create lucrative opportunities for life scientists and traders in health products and resources. Thus, international collaborations can open up transnational spaces that enable innovative research and experimental therapies to thrive. Differences in conditions in life science research and therapies, then, can be a propelling force behind innovation and economic investment, paradoxically stimulating both fierce international competition and international collaborations.

Regulatory and Bioethical Vacuums

Most scientists interviewed in Asia in the present study proved to have explanations for why they, but usually *other* scientists, take considerable risks to further scientific knowledge, treat patients and resist dominant ethical guidelines. Interviews of hundreds of stem cell scientists revealed that all expressed the view that it is professionally important to be led by ethical values. Indeed, most of these scientists stake their reputation on the ability to vouch for the ethicality of their research. Nevertheless, some political regimes do not have compelling ethical regulation and oversight (if only in the shape of 'soft' regulation or social control). Such a regulatory vacuum can be identified because some countries dominate the definition of what bioethical standards are and how they need to be adhered to. A 'bioethical vacuum' would at the same time give stem cell scientists an unfair advantage and saddle them with a reputation as rogue adventurers (see Bharadwaj and Glasner, 2008; Sleeboom-Faulkner and Patra, 2008). But adherence to bioethical guidelines both varies between (Petryna, 2005) and within countries (Sleeboom-Faulkner, 2010b). A bioethical vacuum, therefore, cannot be adequately explained by referring to the absence of morality or a general resistance against hegemonic values of the globally dominant. Without tracing and examining the specific relations between life scientists, policy-makers and patients, and comparing their motives and opportunities to engage in scientific research in high- and low-income countries, it is not possible to gain a clear understanding of the clashes and compromises among scientists and between emerging life values accompanying biotechnology applications in various parts of the world.

National and Global Risk Signatures

Chapter 6 deployed the concepts of national and global risk signature in analysing bioethical considerations from the point of view of stem cell scientists in countries with different institutional and regulatory arrangements, and therefore opportunities. The concepts of national and global risk signature shed light on the relations between actors (i.e. life scientists), institutions (i.e. scientific knowledge, regulation, bioethics committees) and 'things' (i.e. biomaterials, infrastructure, funding), and their potential in cross-border collaborations from the point of view of formal bioethics and in the context of life assemblages. The concept of a global risk signature helps in understanding scientists' views and behaviour in a global context, building on the concept of 'risk signature' developed by Tom Horlick-Jones (2007). The notion of risk signature refers to the risks associated with particular technologies in their specific and concrete material setting, while the concept of global risk signature broadens the application of this concept by integrating it with the international self-awareness of scientists about the differences between national conditions (risk signatures). The conceptualisation of international awareness enables an understanding of the relativity of the perception of risk, rational calculation and ethics, without losing sight of the concrete material relevance of risk-taking in the life science activities and decision-making of scientists. Thus, scientists that hope to take advantage of foreign funding or connection networks are aware of the global risk signatures of bioethics: the awareness that ethical research is important to science institutions at locations where science institutions are highly advanced and well endowed with funding. Awareness of national risk signatures, then, can make the institutionalisation of bioethics a valuable asset to life science centres in low- and middle-income countries, even if the adopted bioethics does not match the broader ethical concerns and rationales of the life assemblage it is part of.

Latent Collaborations

The activities and effort put into living up to the expectations of attractive potential partners by adopting ethical and regulatory aspects of research, I refer to as 'latent collaboration' (Sleeboom-Faulkner, 2013a). Latent collaboration takes place when science institutes adopt measures to instate research ethics to make the institute attractive as a collaborative partner, without necessarily becoming one. At the same time, owing to the costs and dislocation of values inherent in novel concepts, the ways in which such research ethics are integrated and applied develop a character in agreement with their new environment. As the word latent implies, researchers looking for collaborations for strategic reasons are not just passive receptacles of foreign research regulation and practices. They are proactively engaged in the research institutions of countries that lead the intellectual field globally, and keep up with internally acknowledged conventions. While research collaborations can contribute to the scientific goals that various partners hope to achieve through the pooling of resources, Chapter 7 illustrated how they can also be based on unequal partnerships, but provide all with comparative advantage: all partners gain, but some more than others. Such collaborations do not necessarily serve a set of common scientific goals.

Bionetworking

The managerial activities involved in the production of science may be better understood as forms of what Chapter 7 explained as bionetworking. Science collaborations are usually based upon the premise that two or more parties pool research resources, skills and ideas to achieve a scientific goal. In contrast, bionetworking involves co-operation for a range of purposes. Apart from scientific knowledge, this includes the pursuit of honour, patents, access to resources, market dominance and profit. Most life science collaborations have at least some scientific purpose and most contain non-scientific aspects. Some of the non-scientific aspects involve ethics, charity and serving the public, while some of the scientific aspects involve investments that were never meant to finance them. Mapping international collaborations and bionetworking activities by tracing relations between actors, things and institutions may help us understand some of the drivers behind international collaboration and give us clues as to what and whose purposes it will serve.

Life Assemblages

The notion of life assemblages roots bioethics institutions firmly in the realms of the political economy of research, health and morality, incorporating the shifting relations between commercial interests and moral drivers propelling the life sciences. Life assemblages are connections of human and non-human elements that underlie ethical decision-making, be it in formal ethics or everyday life. The notion of a life assemblage subsumes diverse elements—actors, institutions, knowledge, values, facts, practices, and socio-economic, cultural and political resources—that condition and motivate the creation and application of bioethical guidelines and policies.

The generally stated rationale for bioethics institutions is the protection of human life and the prevention of harm and injustice in support of the global moral economy of health and life science research, with the broad aim of finding medical solutions for human suffering on a molecular level. In practice, however, such solutions do not necessarily materialise. The chapters in this book, through a constellation of social-science issues related to population health, biopower, biobanking, public debate, risk, human experimentation, inequality and stigma, related how and why bioethics institutions and research fail to work as envisaged. The study of life assemblages explains that a fundamental problem exists around the financial and health aims of the life sciences due to the diverging life values and levels of wealth of the environments in which life science research and applications develop.

Although problems relating to the friction between the commercial aims and the health aims of life science research are, at least partly, meant to be addressed through the institutionalisation of bioethics, public health institutions and research regulation, at the same time this approach has led to the creation of a new problem cluster in its own right: the mismatch between bioethics and broader life ethics. This mismatch results for contradictory reasons: on the one hand, some societies feel that they need to import alien bioethics institutions lock, stock and barrel, which has led to a dislocation of values; on the other hand, societies that have chosen to take seriously the ethical concerns of members of their population have created bioethics institutions that scientists say puts them at a disadvantage. In turn, the resultant differences in national research conditions tempt scientists, companies and patients to look across borders to establish new connections or collaborations that may involve practices recognised in some areas, but not in others, or, if unsupervised, nowhere at all. These potential problems do not just affect social groups often regarded as society's most vulnerable (e.g. the poor, elderly, disable people, women), but also scientists and medical professionals, who when taking risks to save patients may be ostracised as irresponsible, unethical or quack. In short, the formal bioethics institutions, which have been set up to protect shared life values, health and professional interests of scientists, patients and the public in general, were not just unable to cater for all those concerned, but also created an additional set of problems relating to divergent socio-economic and cultural environments. Some of these problems have been studied in this book, and are captured by the concept of life assemblage.

The research into various dimensions of life assemblages as represented in the chapters of this book has resulted in a number of questions crucial to the co-production of the life sciences and society that takes place in multiple, diverging environments. It became clear that a central question to understanding bioethics in a particular territory is to observe how activities, connections, authorities, pressures and resources that make up life assemblages condition its development. This question required us to clarify how the various aspects of authority, power, governance and decision-making constitute a life assemblage in practice. It appeared that the following aspects of life assemblages are crucial to the translation and embedding of bioethics in networks of power and at various levels of socio-political organisation:

a. Socio-political organisation, political decision-making
b. The receptiveness of the population to biotechnological innovation
c. Pressure to apply 'global' bioethical standards
d. The framing of bioethical problems
e. The role of scientists and experts in public deliberation
f. The moral positioning of bioethics.

The question of how life assemblages relate to bioethics is discussed in relation to the previous chapters of this book.

a. Socio-political Organisation, Political Decision-making

Whether issues belong to the realm of ethics in official discourses and how it is applied depends on the socio-political organisation of society, whose points of view count, and the conditions under which ethical norms and values are formulated, that is, amidst conflict about rights, ownership, culture and religion, or during war, poverty and competition. In the discussion in Chapter 2 on eugenics in the contexts of population policies and reproduction in China, India and Japan, various levels of socio-political organisation turned out to be relevant to the creation of reproductive meaning: the state and provincial organs in policy-making; social movements and non-governmental organisations in lobbying; the clan lineage in the village; the family, the couple and the individual in household matters. Together, these constitute multiple organising bodies with disparate drives and forms of agency over time, and their significance in the conceptualisation and analysis of life assemblages should be weighed accordingly.

State attitudes toward reproducing the population differ starkly. We saw that China, Japan and India have each experienced 'eugenic' reproductive policies. But whereas authorities in Japan in the 1990s rejected any ideology related to 'eugenics', in China and India 'eugenics' had more positive connotations in formal policy-making. It could be argued that official discourses in China and India emphasised the eugenic quality of the population because policies to limit the quantity of the population augmented desires for healthy offspring. However, the notion of eugenics in China is no longer just a matter of political discourse, for it has gained new meanings in the pragmatic attitudes of social groups towards undesirable human traits related to gender and disability. In both China and India gender attitudes surrounding birth have led to 'son preference', a practice prohibited by both governments. Technologies intended to diagnose 'impairment' and policies intended to help small families have been adopted by users for their own purposes. The official ethical requirement in both China and India asking women to give free and voluntary informed consent when reproductive decisions are made is often overridden by the wishes of family households, and by the authority and advice of medical professionals. In contrast, among women in Japan, a society with a higher welfare level and a high standard of medical care, a relatively large number of couples choose to raise a child with a disablement. Whereas in Japan women said they do not want to be subjected to directive counselling, in China and India many families actually asked the doctor to mediate in the decision-making process.

The cases discussed in Chapter 2 illustrate that between the state and individuals there may exist multiple socio-political units, such as the province, town, family household and household lineage, which, to varying extents, modulate official policies. This kind of mediation can take the form of the redistribution of medical resources, the re-interpretation of state policies and also the adaptation of new technologies, the applications of which are marked for gender, seniority and lineage. The study of life assemblages in the context of the politics of reproductive decision-making illustrates that overt bioethics and medical institutions are not just shaped and implemented by nation-state governing institutions, but are mediated by lower-level organisational bodies and adapted differently to multiple levels of organisation at which resources, knowledge, beliefs and policies can follow a relatively autonomous dynamic.

b. The Receptiveness of the Population to Biotechnological Innovation

The urgency with which bioethics institutions have been established has resulted partly from the expectation that the new life sciences will lead to desired health interventions in society. In Foucauldian discourses of biopower it is assumed that the governmentality of nation-state regimes develops along lines conducive to increasing the productivity of the population. Accordingly, new modes of bioethical governance would facilitate the subjectification of individuals, encouraging them to assume responsibility for their health (biocitizenship) by making free, responsible and informed health and reproductive choices. Such subjectification would augment the ability of a population to increase its productivity, and bioethics institutions would help facilitate this process by ensuring that individuals are aware of their prospects and possibilities when confronted with potential medical and biotechnology interventions. Nevertheless, the study of life assemblages in this book shows that the reception of potential biotechnology interventions varies from country to country, and in some cases even points to the opposite trend: the rejection of some new biotechnology applications for ideological reasons.

As discussed in Chapter 3, Japan is a country with well-developed modes of governance, 'biobureaucratic institutions' and 'vertical communication' between layers of organisation in social healthcare. With a population aware of health issues, we would expect Japan to welcome reproductive intervention using predictive technologies. Despite the widespread awareness of political discourses that state the need to give birth to more children in this society characterised by ultra-low fertility, we see neither a growth in family size (Jones et al., 2009) nor clear signs that genetic and prenatal tests are demanded when there is no clear medical indication. Nevertheless, reproductive technologies are not rejected per se. There is a general preference for giving birth in a 'natural', rather than in a 'cold', 'calculated', manner (Lock, 1998; Tsuge, 2008). A similar ontological reservation exists about putting technological perfection above humans. Thus, when we examine women's narratives of giving birth to children with a disablement, we find discomfort with the 'rational' rejection of affected fetuses as potentially handicapped people. This so-called 'eugenic' issue has been a topic of heated debate and underpins ideas and actions shaping life assemblages in Japan.

In contrast, as a result of mainland China's population policy of 'new eugenics' from 1994 to 2003 (Case 1) premarital testing for disorders became widespread. However, as soon as premarital testing was made voluntary, the number of people willing to have a premarital test dropped steeply, and groups that have a genetic propensity for disorders such as thalassaemia and carriers of, for instance, hepatitis and sexually-transmitted diseases were difficult to persuade to go for a test. It is clear that the period of compulsory testing has made candidates aware, not just of the chances of becoming ill or giving birth to affected offspring, but also of the risks of positive testing results to one's chance in the marriage market and its implications for family reputation. In such cases, predictive testing technologies may be avoided out of the rational considerations reigning in this particular life assemblage.

In Case 5, we also saw that under the one-child policy in mainland China, which allows a couple to have another child when the first one is affected by a listed syndrome, the take-up of genetic testing for thalassaemia in coastal areas has increased. In these cases, many parents try to have a second child as a 'saviour' sibling to help the affected child. But as the sibling fetus does not necessarily test as a 'match', it may have to be aborted. Furthermore, the diagnosis is not entirely fool-proof, so that the birth of a second affected child is not rare. One can ask how we can judge what is 'rational' behaviour when complex political and emotional dimensions seem to preclude any definition of rational judgement in the first place. Similarly, those tested positive for Duchenne muscular dystrophy are vulnerable to stigma. As carriers are female and the affected children are male, gender selection could help a couple to give birth to female offspring. But this is problematic: for the daughter might still be a carrier (and will be in a similar predicament when looking for a spouse), while sons are often preferred. The high chance of giving birth to a carrier daughter or an affected son, and the stigma attached to inheritable disease can render entire families ostracised. In such an environment, rational choice rubs against sensitive issues of power.

Among Indian tribes tested for sickle cell disease (Case 4), rational decision-making does not flow from the new set of choices presented to the subjects as a result of the screening technology applied. Instead, it is the meaning attributed to the screening that conditions its effects on daily life. The colour cards distributed among populations formally represented a diagnosis: negative, carrier or positive. But, in practice, the diagnosis was understood in terms of the racial and medical consequences of being diagnosed. Those who were diagnosed as positive were subject to ostracisation, discrimination and stigma. Such interpretations were not intended, as screening was meant to diminish the rate of sickle cell disease (SCD) among tribal groups. The effect of genetic testing on its subjects may be very different according to the cultural environment it is applied in. Although knowledge of test results led to stigma and discrimination in both India and Japan, the institutional environment of tribal areas in India allowed screening results to leak (and acquiesced to the demands for test results of third persons), while in Japan the taboo on 'truth-telling' easily leads to the hiding of results. Considering the socially crippling effects of information on people's genetic make-up to marriage prospects, employment and livelihood, the question of to what extent the rationalisation of reproduction renders society 'more productive' arises, if that is the aim of reproductive policy-makers. It turns out that genetic testing, when meant to rationalise life choices and to have a productive effect on society, can only do so when bioethics institutions have their desired effect and when the population is open to life science applications. Chapter 2 showed that there are many ways in which guidelines on privacy, confidentiality, informed consent and genetic counselling may not serve either economic 'rationalisation' or the purposes of life assemblages. The cases examined showed that Foucauldian notions of biopower do not automatically lead to a disciplining of populations conducive to responsible life choices in the light of the rationale of available technologies.

c. Pressure to Apply 'Global' Bioethical Standards

The relevance of bioethics to research and clinical practices does not only depend on whose interests are served by its use in life science research and clinical applications, but also on the embedding of bioethics in the wider society. Take, for instance, genetic biobanks. The usual justification for having bioethical guidelines and internal review boards is to protect human donors through research guidelines for the sampling of human materials. Without functional socio-political mechanisms of public deliberation it is difficult for notions such as autonomous choice, data protection, the right to available healthcare and the protection of privacy to embed in the fabric of society, and the inclination to apply bioethical standards of informed consent and confidentiality of information to genetic sampling practices may not speak for itself. In such situations, two questions arise related to motivating the public to collaborate with life scientists. First: Why would donors donate if the results based on the research of their samples are not going to benefit them through accessible healthcare? Second: Why should samplers take into account bioethical guidelines if they think that donors do not have any significant rights in need of protection in the first place?

Chapter 2 showed that the introduction of bioethical guidelines and institutions both aids and complicates the sampler's predicament: through their association with 'advanced Western' life science and modern technology, bioethical guidelines enjoy an international aura of reliability and exude authority, progress and promise. As shown in Case 4, on genetic screening in tribal areas in India, leaders of tribal populations may put faith in the sampling cause of prestigious scientists, especially if extra benefits are promised. In this case, bioethical guidelines in the form of informed consent forms provided a veneer of ethicality to the sampling efforts, even when procedures are followed only loosely. Following bioethical guidelines to the letter may make it hard for samplers and scientists when working in an environment where notions of human rights are not appreciated and healthcare access is not enjoyed as a common good. Thus, the National Repository of India was challenged because strict international standards made already collected samples invalid. If revealed, this would require collection owners to re-take the samples. Owners, therefore, preferred to stick with their old sample collections.

The intertwining of ethnic and national identities adds to the ethical complexity of human genetic sampling, conditioning the way in which bioethical guidelines are received and implemented. The human body is strongly associated with notions of national identity and security, so that sampling operations conducted by alien powers tend to be viewed with suspicion. This also became clear in Case 8, on genetic biobanking in Indonesia involving a Dutch and an Australian biobanker. The former, as native of a former colonial occupier of Indonesia, was associated with exploitive practices, while the latter, originally an Indonesian citizen, was regarded as a hero in pursuit of Indonesian scientific progress. Racial and national identity boundaries, then, are relevant to the acceptability of biobanking, and, as such, are part of the life assemblage that underpins the acceptability and application of standards of bioethics.

The Taizhou case study on the longitudinal genetic epidemiological cohort study in China (Case 9) also showed the importance of national identity, but in a different manner. The initial grand claims made about the projected achievements of the Taizhou biobank, including the sampling of millions of citizens in a short space of time, were rapidly converted into bioethically correct jargon after a BBC report revealed that Chinese bioethical standards clearly diverged from those expected by the international science community. It is evident that the pressure on the life sciences to adhere to international regulatory conventions includes calls for bioethics oversight. It is clear that national interests and identities, and international reputation and standards are both important dimensions of life assemblages. It may seem that an emphasis on the differences between life assemblages indicates that members of a life assemblage are inward-looking. But case studies showed that the existence of a life assemblage can only be appreciated through the perception of international pressures and influences as entering from the outside, thus defining (changing) boundaries between in- and outside, and condition of their existence.

The above cases show that national identity can muster support from a national audience for human sample collections, and that the adoption of bioethics guidelines can do the same. But carrying out guidelines may frustrate biobanking ventures, unless notions of collective responsibility and national progress can incite individuals to 'altruistically' donate samples, even when it is likely it will not be of advantage to them or other citizens. As will be discussed below, the appraisal of particular research and bioethical regulation in the context of a 'common good' in tune with the life assemblage may benefit its local embedding.

d. The Framing of Bioethical Problems

Notions of national identity and scientific convention in the regulation of the life sciences lead us to the role of how state politics defines what is an ethical problem. The questions of what prompts a national governing body to recognise a particular issue, and whose problems official bioethics institutions deal with are crucial to the formulation of bioethical guidelines. The discussion around the bioethics of human embryonic stem cell research (hESR) in China and Japan in Cases 10 and 11 showed that dominant bioethics discourses largely associate bioethics with 'Western neoliberalism'. Crudely put, critical opinions of this view regard it as the lubricant of economic exploitation, while nativist views claim that it cannot represent indigenous values.

Despite international constraints and pressure, in practice policy-makers and intellectuals play a decisive role in the interpretation of what are internationally recognised as bioethical problems, and the identification and construction of bioethical problems at home. Through what I here refer to as 'problematisations', they selectively translate internationally acknowledged bioethical problems into stances understood in home-grown paradigms. The use of bioethical regulation for hESR in China and Japan, for example, is justified variously as necessary to compete in a neoliberal world; as necessary to protect citizens and scientific progress; as forced upon by 'the West', hampering otherwise promising scientific progress; and as necessary in terms of native bioethical views on hESR. Intellectuals and policy-makers weave the various elements into what they regard as acceptable bioethics. A case in point is that of hESR in China (Case 10): bioethics was fitted into China's strategic emphasis on stimulating progress in cutting-edge life science research. This dominant view emphasises the 'tradition' of not regarding the embryo as life worth protection and adopting the international standard of 14 days for allowing somatic cell nuclear transfer research, rather than adopting the other, less often mentioned, 'tradition' of viewing the embryo as valuable potential human life (Nie, 2005).

The way in which bioethics is developed, that is, whether it is more or less translated from bioethical guidelines abroad, and whether and how the local population is engaged in deliberations, may be crucial to the stability and workability of a country's bioethics. Stem cell scientists in China compare practices of hESR mainly with those of scientists in what are regarded as advanced countries, while they set them apart from the 'backward' practices of those at home who do not follow 'scientific' standards, including bioethical guidelines. This more outward-looking way of thinking can be contrasted with the more 'inward-looking' views of stem cell scientists in Japan, who, for a long time, had to grapple more with the views of domestic stakeholders than with international standards.

Countries have political mechanisms for dealing with new problems and solutions—'problematisation'—depending on how they define the involvement of their home population in policy-making and may include various forms of public engagement, where mass media, academics, politicians, the Internet and social movements can play an active part. A distinction can be made between regimes that mainly like to know what their subjects do, and those that also like to incorporate their subjects' views in their policies. Thus, although the views of stem cell scientists and ethicists are consulted, Chinese policies have refrained from encouraging broad public debate on bioethics on hESR; the Japanese government, however, has taken initiatives to engage the broader public in debate. Although this policy was only effective in mobilising relatively small groups of patients and activists, it did take on board their views in formulating guidelines on hESR. These diverging political attitudes towards public engagement strongly influence the embedding of bioethics in the science community.

e. The Role of Scientists and Experts in Public Deliberation

The success of public engagement in scientific policy-making is hard to estimate, as it depends on the ways in which existing traditions of public consultation define success. Besides the role of the government discussed above, existing attitudes of scientists towards public debate in general, their ability to engage the public, and their views on the particular aim of the debate largely determine the extent to which the public partakes. An often forgotten effect of public debate is that on scientists. For the existence of a critically engaged public enables scientists to imagine their reaction when trying to work out how to deal with the societal effects of the design of new science projects. Without public deliberation scientists miss a chance to work on what is desirable to a broader public, which disadvantages society in the long term. A governing body could remedy this by organising modes of engagement suitable for the kind of science and technology developments desired in society.

In China, however, the majority of 60 interviewed stem cell scientists expressed objections against public engagement. Having played an active role in the institutionalisation of internal review boards and rudimentary bioethical guidelines, a vast majority of scientists did not welcome broad public debate, as they feared it would turn against their research. In Japan, however, the majority of stem cell scientists were strongly in favour of public debate. Although not particularly keen to engage in debate with staunch critics, stem cell scientists were optimistic about their ability to persuade the 'man on the street' of the usefulness of their research and gain support for its facilitation. Their attitude seemed to pay off. Initially, research guidelines were strict and perceived as obstructions to hESR. When media attention was given to the possibility that Japan could lose its vanguard position to international competition, scientists had available a clear public history of their cause. This facilitated the decision to adjust guidelines and increase investment in stem cell science in 2009. To a large extent, therefore, in Japan, debate justified life science research and an attempt was made to make it acceptable to the population, while in China public debate was perceived as a risk—the risk of making science controversial. These examples show how bioethics is attuned to, but often in contradiction with, the life assemblage it is part of.

Although the state plays an important role in organising public debate, the engagement of the population in hESR in Japan and South Korea was not just a matter of the government stimulating debate. Both countries harbour organisations and social and religious movements that are free to make public their views on, for instance, the status of the embryo, patients, feminism and disability. The views of these 'minorities' were influential in the establishment of bioethics committees, regulatory committees and forms of governance adopted in hospitals, laboratories, companies and in medical associations. In fact, the debate underpinned the formulation of bioethical guidelines. But as bioethics has increasingly become a matter of professional expertise, it has become difficult for civic groups to keep up to date with developments and to invest time, money and effort into the activities required to mobilise the views of citizens. The extent to which bioethics is being professionalised and institutionalised in a country is therefore important to how the public is being incorporated in the creation of bioethical guidelines, and is also of influence on the way bioethics is formulated.

f. The Transnational Awareness of Scientists

The transnational perspectives of scientists are part of the life assemblage they work in, shaping attitudes to scientific research and bioethics. The ability to compare across borders makes the very rules of bioethics expedient to both competition and collaboration between scientists and medical professionals in countries with different regulatory, wealth and health environments. In fact, an awareness of the material and regulatory conditions for conducting research may be part-and-parcel of the strategic knowledge scientists have when deciding to engage in international collaboration. The formulation of bioethics partly depends on the ability of scientists to form networks abroad and on their awareness of comparative advantage in various fields. In stem cell therapies, such comparative advantage may include access to patients, favourable regulation, funding availability, affordable expertise, equipment and science knowledge.

Reasons for initiating international collaborations vary from enabling the achievement of a scientific aim to merely making profit. Most collaborations combine advantages, but when a collaboration thrives owing to conditions abhorred in one country or by systematically producing conditions that promote poverty and inequality, such life science collaboration is best characterised in terms of bionetworking activities largely motivated by profit. Trying out experimental therapies in other countries where patients are available and therapies can be tested under conditions less stringent, oversight takes advantage of inequalities and is better understood as bionetworking than as international science collaboration. The problems associated with the strategic use of variations in life assemblages are new and convoluted, and often depend on an international audience much less aware of the complications than the scientists involved. Under such international circumstances the playing out of difference in life assemblages also effects change in life assemblages, generating new forms of 'international bioethics' that are heavily influenced by transnational factors that people at home are hardly aware of. This convolution challenges the notion of the standardisation of bioethical guidelines as it can facilitate locally undesirable practices when bioethical guidelines are not implemented as intended, and as a lack of reliable bioethical capital can hamper scientific progress in the eyes of well-funded elite scientists who fear their reputation will suffer if they are associated with less bioethically endowed institutions at home. Both the view of bioethics as facilitating exploitation and as hampering scientific progress has led to bioethical subversion. Thus, scientists that have been criticised for their experimental exigencies have gone 'underground' as private practitioners in the branches and offices of large public and private healthcare organisations (Song, 2011), while bioethics regulation in wealthy societies is feared to have driven away innovative scientists to countries without ethical scruples (Sleeboom-Faulkner and Patra, 2008). It is clear that a life assemblage conditions scientists' reflection on bioethics regulation and their resistance to it, the consequences of which are, in turn, an important component of it.

'Research Objects' and the 'Common Good'

Critics have referred to the adoption of bioethics institutions to enable life science research without regard for the particular environment in which it is practised as the 'dislocation of bioethics' (Bharadwaj and Glasner, 2008; Sleeboom-Faulkner and Patra, 2008). The question of what it means for a population when its government decides to pursue advanced life sciences is not always asked, despite the fact that the population's co-operation, investments and engagement are essential to the wide appreciation of the life sciences in society in the long term. To ask how investment into particular science innovations (through donation, education and taxes) renders benefits to the people who indirectly support research is important for the acceptability of new biomedical applications. Discussions on the suitability of forms of public participation have especially concentrated on how the public can be included in science policy decision-making in a fair manner without unnecessarily obstructing science innovation. The increasing standardisation and professionalisation of bioethics affects life assemblages variously through different forms of public involvement in debate on science bioethics in different locations.

Common Good

Questions about the institutionalisation of democratic procedure and human rights have received much attention in the study of politics, while considerations of what activities constitute the material and political dimensions of life science research practice, such as research funding, regulation, infrastructure and sourcing, and how they affect the people and institutions in political constituencies have received less prominence. These issues are of great relevance to the way a country spends its financial budget on healthcare, research and welfare for its population, the way it is prepared to support companies, scientists and citizens with infrastructure, laws and regulation, and eventually influence the motivation of scientists to conduct particular kinds of research (Sleeboom-Faulkner and Hwang, 2012).

It is these material and political aspects that especially matter to a limited number of people involved in a particular form of life science or biotechnological application. They can include core groups of scientists, donors, patients, and those with particular ideological and religious causes. Similarly, the drive for scientific progress and competition in the life sciences has gained more attention than the practical issues of how funding, regulation, and material sourcing can best contribute to what people in a certain region or political constituency regard as a 'common good', such as taking care of the sick and elderly. Even when agreement exists about participatory procedures for political decision-making about sensitive research topics such as hESR and genetic testing, when not underpinned by accepted policies on public health and life values, it is hard to justify support for the research. Thus, it is difficult to defend some forms of life science research in a country usurping large amounts of government investment and utilising bodily material donated by volunteers if the research is not aimed at a commonly recognised good, such as providing the country's patients with accessible and affordable healthcare.

The concept of a 'common good' as used here is based on the elaboration of the Aristotelian concept of politics by Gerard de Vries (2007). A research practice that is not driven by a 'common good' corresponding to the values prevalent in the constituency that provides its material and political support, including funding, expertise, education, infrastructure, patients, oocytes and embryos, does not have broad public support. According to de Vries (2007), what has been lacking thus far in discussions about the ethics of scientific research is not so much attention to the political negotiation of authority and legitimacy of bioethical procedures in general, but attention to what he calls the 'object' of research: the driving force behind research, that is, a practice serving a common good. Explaining the concept of 'research object' involves reflection on the political role of scientists and their involvement in the 'common good'. Rather than putting the main emphasis on the issue of who should be involved in debates underlying science decision-making, de Vries shifts the question to what it means for scientists and concerned others to be politically involved. According to de Vries, experts engage in politics when they try to translate a wide range of conflicting views and interests into a 'common good' served by their research, such as 'finding a cure for X by X method to the benefit of X'. Such a 'research object' would involve setting up a constitution, including an organisational structure, institutions, rules and provisions, for a practice in which the 'object' could circulate. Thus, if in the study of genetic testing of SCD we regard the prevention of SCD-affected children (PSCD) as a common good, then we find that not all genetic samplers are engaged in politics in the sense of serving such a 'good'. Instead, some samplers are more interested in career-making, profit-making or establishing relations of mutual benefit. In such cases, PSCD as a research object does not gain currency. As de Vries argues, 'To be in or out of politics...depends on whether an actor is involved in a *praxis* that aims at a political object, or not' (2007: 794–5). In some cases the political object is flawed from the beginning. Thus, some SCD screening programmes, as shown in Case 4, do not offer counselling, medicine or follow-up treatment, so that the screening defeats the object for most of those screened. Although samples are banked and used for research, such health intervention may leave a population worse off than before the intervention. In such a life assemblage the 'common good' does not tally with the bioethics regime. The question arises, then, as to how health intervention and research can be mobilised to serve a common good in congruence with the views of the people supporting, engaged in and benefiting from them. Bioethics could facilitate such a common good: not necessarily by basing it on dominant cultural, political or religious notions, but through debate of reigning life values involved in the case in question.

Standardisation of Bioethics and Professionalisation

A life assemblage covers a broader conceptual field than does bioethics, and its understanding requires us to look beyond bioethics. It leads us to reflect on questions about the way an object is translated into local socio-political concerns, the mode of public deliberation and participation, and the high expectations of the life sciences to deliver cures where the regeneration of life is being commodified. Instead of shaping bioethics reflectively in light of these questions, solutions have been sought in the direction of the global standardisation of bioethics and its institutionalisation.

When official guidelines in countries are created in the light of standards utilised by international science communities, such as those concerning informed consent and the moral status of the embryo as supported by the International Society for Stem Cell Research, these guidelines may be of little interest to local experts and the general public in a country. Cases 10 and 13, on stem cell research in China, and Case 16, on experimental stem cell therapies in India, showed that the implementation of such standards becomes unreliable. In this way, the relevant concerns of the public get missed, and the opportunity to engage society in discussion is lost. To counteract such a reaction, governments and groups of scientists have concentrated on strengthening capacity-building in bioethics and setting up ethics committees led by professionals, including doctors, lawyers, and social science and humanities scholars. Even though such initiatives have had an effect in elite laboratories and hospitals, the standardisation and professionalisation of bioethics have also led to problematic scenarios where a corresponding 'research object' is lacking. In such a case, research does not serve a common good.

Standardisation, manifest in the institutionalisation and professionalisation of bioethics, may, in fact, discourage debates on whether to support, for instance, hESR, who should benefit from the fruits of hESR, how this will happen, and at whose and what cost. Although these questions are integral to the politics of developing a 'common good' for circulation in a political constituency, when no such object can be identified the practice of hESR becomes an activity governed by other interests. In such cases, standardisation and professionalisation of bioethics become buffers against external critique and debate, protecting standards that bear no relation to the object of hESR. For instance, why should embryos or oocytes be donated to for-profit therapies practised on an experimental basis when no return conducive to the 'common good' can be expected?

In both Japan and South Korea, although religious, patient and feminist groups have been active participants in public debate on hESR, issues relating to the amount of time, effort and money spent on organising meetings, website maintenance and consultation for a sustained engagement in debate have become problematic. With the increase of state investment in training capacity, the increase of bioethics expertise among professionals and the bureaucratisation of guidelines, 'lay experts' have found it difficult to cope with the demands placed upon them; as a result, one of the main groups engaged in public debate in Japan, the Anti-Eugenics Thought Network, disappeared in 2007. The professionalisation of bioethics increasingly means that, apart from the time, effort and funding for activities, activists require legal, scientific and interactional skills to function.

Implications for Research

Where life science research is required to have a research object involving the common good of the people concerned, the frameworks through which scientists, medical professionals and patients make choices about the relevance of research and treatment could be made more in line with the healthcare and research needs of local populations. If research in countries with scarce health resources and low standards of living would be relevant to local healthcare needs, patients may have fewer scruples about collaborating with scientists, and scientists may, rather than regarding a frank public discussion as a risk or threat, welcome the opportunity to persuade the community to support their work. Although existing differences of healthcare provision, science funding, education and regulation will continue to attract chancers, the incentives for scientists to engage in bionetworking activities could well diminish. Further empirical research is needed to shed light on how research in agreement with the 'common good' affects the decision-making of scientists, choices of the population and the development of the life sciences—in other words the life assemblage.

Responsible Innovation

The desirability of bioethical uniformity, implied by the term 'global moral economy', suggests that life science research and applications, with the right form of bioethics institutions, will lead to human health improvement. We have seen, however, that economic and moral drivers clash and intertwine to generate practices that in many circumstances cannot be regarded as either productive or moral. Knowledge of life assemblages provides us with insights into the kinds of larger challenges associated with a particular form of life science innovation, which is useful when initiating discussion on the direction of it and the conditions under which it should take place.

The construction and study of life assemblages showed that bioethics institutions develop through various levels of political organisation, through modes of boundary-making between the 'bioethical self' and 'others', through strategic uses of bioethical regulation, and through 'latent collaboration' and 'bionetworking'. Although bioethical guidelines are justified as ways of dealing with the discrepancy between commercial, research and health aims of life science research, we have seen how bodies of bioethical guidelines are formed variously through locally-driven political, scientific and personal aims, through global pressures of collaboration and competition, and through the push and pull of markets. Capturing the real-world heterogeneity of the things, rules, drives and actors that shape bioethics and research regulation, life assemblages enable us to better understand and recognise relations between bioethical regulation and their local embedding:

- the effects and interactions between centralised and more local forms of regulation, including state regulation of birth planning and scientific research, and its pragmatic adaptation by communities, local households and individuals;

- the internationalisation of bioethical standards and bioethics discourses, and their effect on and interactions with local practices;
- the relationship between national bioethical problematisations and representational ideologies as political constructs of local stakeholders;
- the relationship between technological innovation and the intervention in reproductive practices of concrete communities (i.e. biopower);
- the role of the public in life science policy-making and the concrete role the public plays in the lives of scientists and medical professionals; the ways in which transnational differences and similarities create possibilities for exchange and opportunities for investment, politics and health, and offer incentives to local life science initiatives.

Recognising relations between bioethical regulation and their local embedding as life assemblages may considerably facilitate responsible innovation in a world of local and global inequalities.

The great variety in the forces and stakeholders influencing the development of bioethics institutions raises the question of what kinds of research and biotechnological applications to support in the first place. The identification of a 'research object' relevant to a life assemblage and means available for developing it, and translating it into a widely circulating praxis may be crucial to a successful realisation of a 'common good'. In fact, where there is no clear research object associated with biotechnological applications, bioethics institutions may harbour unrevealed social and political problems. Public discussion, by translating a wide range of conflicting views and interests into a 'common good', might be crucial to the shaping of a constitution for a practice in which stem cell research could circulate or, if unacceptable, may be strategically put aside. If conducive to setting up such a practice, debate may actually be more effective when the range of professionals and lay people engaged with the material and political aspects of biotechnology are allowed and stimulated to develop their views, and are given serious consideration, rather than by creating a 'democratic' system to poll views of all individuals in a particular constituency. Such discussion would need to weigh the costs and current possibilities of scientific applications against the material conditions and political ideals subscribed to in the life assemblage.

The definition of a research object as research based on political involvement serving a 'common good' draws attention to the significance of the following questions in the context of the globalisation of bioethics debate and the life sciences. First, how can a research object that serves a common good in society be identified? After all, it is questionable whether all scientists are prepared to engage in setting up a structure and modes of governance for a research practice that aims to lead to cures for circulation in the constituency that supports their research financially and materially. Second, how do we draw the boundaries of the constituency in which scientific research takes place? This question requires us to examine whether experts use local resources to provide their services as a global 'common good' and how such good would circulate among those it is meant for. The question is similar to asking how research can serve a particular life assemblage. And, last, do intention, means and motivation exist to translate the benefits of a global 'common good' into the practice of the political constituency researchers are part of? This question asks how research facilitated transnationally can serve the 'common good' of the life assemblage concerned, rather than just the aims of the companies and scientists directly involved in collaborative networks. The answers to these questions may help us to determine the role that experts and the public should play in debate on bioethics and the life sciences in an international context.

Notes

1 Introduction: From Global Moral Economy to Assemblages of Life

1. See http://www.cbd.int/
2. http://www2.lse.ac.uk/researchAndExpertise/units/BIONET/; http://www.eubios.info/; http://www.ethox.org.uk/; http://asiapacificstemcells.org/ (last accessed March 2011).

2 Reassembling Populations: Questions of Eugenics in China, India and Japan

1. See http://www.geneticsandsociety.org/policies/international/council.html
2. This is the first sentence of the Eugenic Protection Law (1948–1996) where the purpose of the law is stated. See http://www.lit.osaka-cu.ac.jp/user/tsuchiya/gyoseki/paper/JPN_Eugenics.html (last accessed 6 September 2011).
3. Parts of this case study have appeared earlier in a different, longer version in *The China Journal* (Sleeboom-Faulkner, 2010a).
4. Parts of this case study have appeared earlier in a different, longer version in *The Journal for Bioethical Inquiry* (Gupta, 2007).
5. Parts of this case study have appeared earlier in a different, longer version in *Culture, Health & Sexuality* (Kato, 2009a).
6. Under this law, 47 men and 47 women were sterilised in 1941, 106 and 83 in 1942, 90 and 62 in 1943, and no men and 18 women in 1944 (Kôseishôimukyoku, 1955: 828).
7. Under this law, 16,520 people were sterilised between 1949 and 1997 (Revision of the Eugenic Protection Law: Different standpoints (Asahi Shimbun, on 14 February 1996)).
8. Tsuji describes the transformation and struggles in decision-making about whether to test a fetus during pregnancy following the birth of a child with Down syndrome (2003: 51–3).

3 Biopower and Life Assemblages: Genetic Carrier Testing in India, China and Japan

1. Parts of this case study have appeared in a different, longer version (Patra and Sleeboom-Faulkner, 2010).
2. Parts of this case study have appeared earlier in a different, longer version *Social Science and Medicine* (Sleeboom-Faulkner, 2011a) and in *Culture, Health and Sexuality* (Sui and Sleeboom-Faulkner (2010a).
3. An altered, shortened version has appeared previously in *Biosocieties* (Kato, 2009b).

4. Tamura obtained an official certificate of genetic counsellor in the USA and she practices counselling in Tokyo.
5. Yomiuri Online (2005); *Japan Times Weekly* (2005).
6. For example, in Long (2005), there are a number of stories of interactions among patients, their family members, friends and medical doctors regarding the issue of disclosure of illness (*Kokuchi*).

4 Human Genetic Biobanking and Life Assemblages in Asia: Transnational Moral Economies of Health, Progress and Exploitation

1. Other genetic and archaeological evidence shows that there never was an evolutionary separation between races, and the presumption that a purer race existed in the past is incorrect (see Templeton, 2003).
2. For a long list of declarations against the HGDP, see Macer (1997).
3. Parts of this case study have appeared earlier in a different, longer version (Patra and Sleeboom-Faulkner, 2009a).
4. Parts of this case study have appeared earlier in a different, longer version in *The Asian Social Science Journal* (Sleeboom-Faulkner, 2007).
5. Bioethics-related activities in Indonesia 2005: MOH No 1334/2002, see www.knepk.litbang.depkes.go.id/knepk/download%20dokumen/presentasi (accessed May 2011).
6. Two memoranda of understanding (MOUs) were signed by the Dutch and Indonesian ministers in support of the Scientific Programme Indonesia-Netherlands (SPIN) co-ordinated by The Royal Netherlands Academy of Arts and Sciences (KNAW). Since 1997, infectious diseases and biotechnology constitute two of the six focus points of research (Stapel, 2003).
7. Dr Edison Liu, Executive Director of the Genomics Institute of Singapore (GIS).
8. Dr Jahja Santoso, Apt., President Director of Sanbe Farma.
9. Most Indonesians are agreed that political 'reformasi' since 1998 and the fall of Suharto have failed to check corruption, commonly referred to as KKN. The government of President Susilo Bambang Yudhoyono has made the fight against KKN its chief priority, but the results so far have been disappointing.
10. The Government Regulation of the Republic of Indonesia No. 39 of 1995 concerning health research and development was established on 14 November 1995.
11. Dr X, 1 August 2002, http://www.ihdreg.com/mainpage.htm (last accessed January 2011).
12. Parts of this case study have appeared earlier in a different, longer version in *East Asia Journal for Science and Technology Studies* (Sleeboom-Faulkner, 2013b).
13. Communication with (anonymous) manager of the Biobank in April 2009.
14. Personal communication Poppy Toland (April 2009). Also, see Toland (2007).

5 Life Assemblages of Human Embryonic Stem Cell Research in China and Japan: Bioethical Problematisations and Bioethical Boundary-making

1. Parts of this case study have appeared earlier in a different, longer version in *Biosocieties* (Sleeboom-Faulkner, 2008a).
2. Medical textbooks examined: (1) Li, B., Li, Z. and Cong, Y. (1996) *Yixue Lunlixue* [*Medical Ethics*] (Beijing: Beijing Yike Daxue Chubanshe); (2) Lu, Q. (main ed.) (1999) *Yiue Lunlixue* [*Medical Ethics*] (Huazhong: Keji Daxue Chubanshe); (3) Yang, F., Zhang, C., et al. (2000) *Yiue Lunlixue* [*Medical Ethics*] (Shanghai: Di Er Jun Yi Daxue Chubanshe); (4) Zhang, X. (2000) *Yiue Lunlixue* [*Medical Ethics*] (Beijing: Guofang Keji Daxue Chubanshe); (5) Qiu, X. (main ed.) (2001) *Yiue Lunlixue* [*Medical Ethics*] (Beijing: Renmin Weisheng Chubanshe); (6) Chen, Y., Wang, D., Feng, Z. and Lu, W. (2002) *Dangdai Yiue Lunlixue* [*Contemporary Medical Ethics*] (Beijing: Beijing University Press); (7) Du, J. and Wang, X. (main eds) (2002) *Yiue Lunlixue Jiaocheng* [*A Textbook in Medical Ethics*] (Beijing: Kexue Chubanshe); (8) Bu, P. (main ed.) (2003) *Yiue Lunlixue* [*Medical Ethics*] (Beijing: Gaodeng Jiaoyu Chubanshe); (9) Qiu, X. (main ed.) (2003) *Yiue Lunlixue* [*Medical Ethics*] (Beijing: Renmin Weisheng Chubanshe); (10) Li, R. and Liu, Y. (eds) (2003) *Yiue Lunlixue* [*Medical Ethics*] (Changsha: Zhongnan Daxue Chubanshe); (11) Lu, Q., Ruan, L. and Zou, C. (main eds) (1999) *Yiue Lunlixue* [*Medical Ethics*] (Huazhong: Keji Daxue Chubanshe); (12) Guo, Z. (main ed.) (2004) *Yiue Lunlixue* [*Medical Ethics*] (Xian: Di Si Jun Yi Daxue Chubanshe); (13) Cao, K., Qiu, S. and Fan, M. (2004) *Yixue Lunlixue* [*Medical Ethics*] (Shanghai: Fudan Daxue Chubanshe); (14) Sun, M. (main ed.) (2004) *Yiue Lunlixue* [*Medical Ethics*] (Beijing: Gaodeng Jiaoyu Chubanshe); (15) Li, Y. (2004) *Yixue Lunlixue* [*Medical Ethics*] (Zengzhou: Zengzhou Chubanshe); (16) Che, L. (2005) *Yixue Lunlixue* [*Medical Ethics*] (Beijing: Gaodeng Jiaoyu Chubanshe); (17) Wu, S. (2005) *Yixue Lunlixue* [*Medical Ethics*] (Guangzhou: Guangdong Chubanshe).
3. Parts of this case study have appeared earlier in a different, longer version in *Biosocieties* (Sleeboom-Faulkner, 2008a) in *Science as Culture* (Sleeboom-Faulkner, 2008b).
4. On 6 October 2001, the ethics committee of the Southern hESR Centre published the 'General Ethical Guidelines for hESR' and, in 2002, the expert committee of the Ministry of Health published the 'Proposal for the Supervision of Ethical Principles for hESR' (10: 175; MoH, 2003c). It greatly resembles regulation existing in Europe and the USA. Both sets of guidelines said that research embryos may derive from spare, aborted and cloned embryos. The donation of spare embryos is allowed only on a voluntary basis and with informed consent, with a research window of 14 days. The insertion of the blastocyst into the uterus is forbidden; it should have no use other than for medicine. The co-ordinator of embryo donation and the researcher may not be one and the same person, and the identity and information on the donor must remain private (also see Doering, 2003).
5. The three ministries that issued the guidelines were the Ministry of Education, Sports, Science and Technology (MEXT), the Ministry of Health,

Labor and Welfare (MoHLW), and the Ministry of Economy, Trade and Industry (METI).

6. Lecture by Ida Ryûichi (member of the bioethics committee of the CSTP) on bioethics, law and stem cell research, Kyoto University, 31 April 2006. Professor Ida kindly gave me his PowerPoint presentation.

7. Pure Land Shin Buddhism is a branch of Pure Land Buddhism, which, in turn, is a branch of Mahayana Buddhism. The Pure Land Sect (Jôdo Shinshu) is practised mainly in Japan (Shimazono, 2008).

8. Parts of this case study have appeared earlier in a different, longer version in *East Asia Journal for Science and Technology Studies* (*EASTS*) (Sleeboom-Faulkner, 2010c) and *Medical Anthropology* (Sleeboom-Faulkner, 2010d).

9. It is only rare that a scientist points out the freedom private clinics have to conduct experimental research. It is certainly not part of a common discourse.

6 Scientists and the Public in East Asian Life Assemblages: Risk, Debate and the Professionalisation of Bioethics

1. Parts of this case study have appeared earlier in a different, longer version in *Health, Risk and Society* (Sleeboom-Faulkner, 2010b).

2. Nevertheless, variation exists between Germany and the UK in regard to the strictness with which researchers adhere to regulation (Weber and Wilson-Kovacs, 2008).

3. A cybrid is a hybrid cell that combines the nuclear genome from one source with the mitochondrial genome from another source (http://en.wikipedia.org/wiki/Cybrids_(medical)).

4. Parts of this case study have appeared differently in part of an earlier article published in *Social Studies of Science* (Sleeboom-Faulkner and Hwang, 2012).

5. Parts of this case study have appeared earlier in a different, longer version in *New Genetics and Society* (Sleeboom-Faulkner, 2011b).

6. Similar views were obtained by Tsuge (2008); also see Lock (1998).

7. MEXT revised its guidance for hESR (effective from 21 August 2009) by omitting the review by MEXT and easing the regulation for transferring cells created from human embryonic stem cells among research institutions (see MEXT, 2009).

8. Even the wider public in support of iPS desired international regulation and restraint (Shineha et al., 2010).

7 Life Assemblages and Bionetworking: Developments in Experimental Stem Cell Therapies in India and Japan

1. Parts of this case study have appeared earlier in a different, longer version in *Anthropology & Medicine* (Patra and Sleeboom-Faulkner, 2009b).

2. These data were shared during a personal communication with Dr Sujata Mohanty, Assistant Professor, ORBO and Stem Cell facility, All India Institute of Medical Sciences, New Delhi, with PKP on 26 November 2008.

3. See http://living.oneindia.in/insync/stemcelltherapy-260207.html (last accessed on 14 January 2009).

4. Another pseudonym for a stem cell company attached to the Y Group of Hospitals, which has a branch office in Malaysia.
5. Parts of this case study have appeared earlier in a different, longer version in *Social Studies of Science* (Sleeboom-Faulkner and Patra, 2011).
6. ZCRM has research collaborations with many small Indian institutes and hospitals: the Vision Research Foundation, Sankara Nethralaya, Chennai; The Centre for Liver Research & Diagnostics, Owaisi Hospital, Hyderabad; the Aditya Jyot Eye Hospital, Mumbai; the Institute of Ophthalmology, Joseph Eye Hospital, Trichy Cell Therapeutics; the Department of Sports Medicine, SRMC & RI (DU), Chennai; the Department of Molecular Reproduction, Development & Genetics, Indian Institute of Science, Bangalore; the Institute of Pathology, Indian Council for Medical Research (ICMR), Safdarjung Hospital Complex, New Delhi; and Sahajanad Laser Technologies, Ahmedabad.
7. See Chapter 6.
8. Using cytotoxic T lymphocyte cells is expensive, as it requires special equipment. According to the developers of the therapy, It Is disputed whether the use of natural killer cells alone is sufficient as they increasing the chance of developing infection.
9. In Japan, private clinics can provide any therapy as long as the patient consents. But for most scientists this is not an option as such experimental therapies are not covered by insurance. Moreover, scientists put their reputation (and funding) at risk when they engage in dubious practices. It is hope that data garnered from trials of therapies in India will facilitate the authorisation of clinical trials in Japan.

References

Abdullah, I. and Hüsken, F. (2003) 'Social Security Research in a Time of Crisis', in Stapel, J. (ed.) *Scientific Programme Indonesia–Netherlands (SPIN)*, pp. 17–26 (Amsterdam: KNAW).

Acharya, T., Kumar, N. K., Muthuswamy, V., et al. (2004) '"Harnessing Genomics to Improve Health in India"—An Executive Course to Support Genomics Policy', *Health Research Policy and Systems*, 2, 1.

Adam, B. (2007) 'Introduction: Repositioning Risk: The Challenge for Social Theory', in Adam, B., Beck, U. and van Loon, J. (eds) *The Risk Society and Beyond*, pp. 1–31 (London; Thousand Oaks; New Delhi: SAGE).

Adams, V., Erwin, K. and Le, P. V. (2009) 'Public Health Works: Blood Donation in Urban China', *Social Science & Medicine*, 68, 3, 410–18.

Agamben, G. (1998) *Homo Sacer: Sovereign Power and Bare Life* (Stanford, CA: Stanford University Press).

Alford, P. (2008) 'Japan's Stem Cell Research Surge', *The Australian*, 19 Jan.

Angastiniotis, M., Modell, B., Englezos, P. and Boulyzhenkov, V. (1995) 'Prevention and Control of Hemoglobinopathies', *Bulletin of the World Health Organization*, 73, 375–86.

Anon (2002) 'Bioinformatics—India Discovers Goldmine', available at: www.indian purchase.com/magonline/purchase/200208/article1.htm (accessed 29 December 2005).

Anon (2005a) 'Birth Defects on Rise as China Loosens Premarital Checks', *People's Daily*, 3 Mar.

Anon (2005b) 'China Considers Re-imposing Premarital Health Check-up System', *People's Daily Online*, 1 Sep, available at: http://english.peopledaily. com.cn/200509/01/eng20050901_205857.html (accessed 29 April 2010).

Anon (2007) 'Birth Defects on Rise as Checkups Slide', *China Daily*, 21 May.

Anon (2008) 'Japanese Scientist Says Regulations Needed for Non-embryo Stem Cells to Avoid Abuse', *The International Herald Tribune*, 9 Jan.

AnSI (Anthropological Survey of India) (2006) Official Website of the Anthropological Survey of India, Government of India, Kolkata, available at: http://www.ansi.gov.in/policy_interventions.htm (accessed 22 September 2010).

Ashcroft, R. E. (2010) 'Cold Human Rights Supersede Bioethics?' *Human Rights Law Review*, 10, 4, 639–60.

Bak, H.-J. (2007) 'Perceptions and Evaluations of Norms of Science among Korean Scientific Community', *Science and Technology Studies*, 7, 2, 91–124.

Basu, S. K. (ed.) (1994) *Tribal Health in India* (New Delhi: Manak Publication).

Baumann, G. (2002) *Contesting Culture. Discourses of Identity in Multi-ethnic London* (Cambridge: Cambridge Studies in Social and Cultural Anthropology).

Baylis, F. (2009) 'The HFEA Public Consultation Process on Hybrids and Chimeras', *The Kennedy Institute of Ethics Journal*, 19, 1, 43.

Beck, U. (1992) *Risk Society: Towards a New Modernity* (London: SAGE).

Beck, U. (2007) [1999] *World Risk Society* (Cambridge, MA: Polity Press).

Bender, W., Hauskeller, C. and Manzei, A. (eds) (2005) *Grenzüberschreitungen* [*Crossing Borders*] (Münster: Agenda Verlag).

Bergstrom, P. (ed.) (2004) *Ethics in Asia-Pacific* (Bangkok: UNESCO).

Bharadwaj, A. and Glasner, P. (2008) *Local Cells, Global Science. The Rise of Embryonic Stem Cell Research in India* (London; New York: Routledge).

Birch, K. and Tyfield, D. (2013) 'Theorizing the Bioeconomy: Biovalue, Biocapital, Bioeconomics or...What? *Science, Technology and Human Values*, 38, 3, 299–327.

Bloom, G., Kanjilal, B. and Peters, D. H. (2008) 'Regulating Health Care Markets in China and India', *Reform Goes Global*, 27, 4, 352–63.

Blumenthal, D. and Hsiao, W. (2005) Privatization and its discontents – the evolving Chinese health care system, *The New England Journal of Medicine*, 353, 1165–70.

Bonnicksen, A. L. (2002) *Crafting a Cloning Policy: From Dolly to Stem Cells* (Washington, DC: Georgetown University Press).

Brandt-Rauf, S., Raveis, V., Drummond, N. and Rothman, S. (2006) 'Ashkenazi Jews and Breast Cancer: The Consequences of Linking Ethnic Identity to Genetic Disease', *American Journal of Public Health*, 96, 11, 1979–88.

Callon, M. (1998) 'Introduction: The Embeddedness of Economic Markets in Economics' in Callon, M. (ed.) *The Laws of the Markets*, pp. 1–57 (Oxford: Blackwell).

Cao, D. (2004) 'Premarital Check-up Vital to Family Health", *China Daily*, 30 Nov.

Cavalli-Sforza, L. L., Wilson, A. C., Cantor, C. R., Cook-Deegan, R. M., and King, M. C. (1991) 'Call for a Worldwide Survey of Human Genetic Diversity: A Vanishing Opportunity for the Human Genome Project', *Genomics*, 11, 490–1.

CDST (Center for Democracy in Science and Technology) (ed.) *Paradox of Progress: Toward Democratization of Science and Technology* (Seoul: Dangdae).

Chakrabarty, D. (1997) 'Postcoloniality and the Artifice of History: Who Speaks for "Indian" Pasts?', in Guha, R. (ed.) *A Subaltern Studies Reader 1986–1995*, pp. 263–93 (Minneapolis, MN: University of Minnesota Press).

Chan, T. Y. and Critchley, J. A. (1996) 'Usage and Adverse Effects of Chinese Herbal Medicine', *Human & Experimental Toxicology*, 15, 5–12.

Chen, X. G. (2005) 'Beijing Cheng Zhi Xujia Yilaio Guanggao' ['Beijing Punishes Exaggerated Medical Advertisement]', *Zhonghua Gongshang Bao* [Chinese Industry and Commerce Daily], 10 Aug.

Chen, Z., Chen, R., Qiu, R., Du, W. and Lo, Y. (1999) 'Chinese Geneticists are Far From Eugenic Movement', *American Journal for Human Genetics*, 64, 1199.

Chu, J., Huang, W., Kuang, S. Q., et al. (1998) 'Genetic Relationship of Populations in China', *Proceedings of the National Academy of Science USA*, 95, 20, 11763–8.

Cooper, J. (1990) *Chinese Alchemy: the Daoist Quest for Immortality* (London: Sterling Publishing).

Cooper, M. (2008) *Life as Surplus. Biotechnology & Capitalism in the Neoliberal Era* (Seattle, WA; London: University of Washington Press).

Cooper, M. (2010) 'Turbulent Worlds: Financial Markets and Environmental Crisis', *Theory Culture & Society*, 27, 2–3, 167–90.

CRCF (Chinese Red Cross Foundation) (2005) ['Fund Helps 650,000 Children with DMD'], available at: http://www.crcf.org.cn/sys/html/lm_4/2007-09-12/113730.htm (accessed 3 December 2013).

Critchley, J. A., Zhang, Y., Suthisisang, C. C., Chan, T. Y. and Tomlinson, B. (2000) 'Alternative Therapies and Medical Science: Designing Clinical Trials of

Alternative/Complementary Medicines, is Evidence-based Traditional Chinese Medicine Attainable?', *Journal of Clinical Pharmacology*, 40, 462–67.

Cyranoski, D. (2004) 'Korea's Stem Cell Stars Dogged by Suspicion of Ethical Breach', *Nature* 429, 3.

Cyranoski, D. (2005) 'Japan's Embryo Experts Beg for Faster Ethical Reviews,' *Nature*, 438, 7066, 263.

Cyranoski, D. (2008) 'Stem Cells: A National Project', *Nature*, 451, 7176, 229.

Daley, G. O., Ahrlund Richter, L., Auerbach, J. M., et al. (2007) 'The ISSCR Guidelines for Human Embryonic Stem Cell Research', *Science*, 315, 603–4.

DBT (Department of Biotechnology, Government of India) (2002) 'Ethical Policies on Human Genome, Genetic Research and Services', available at http://dbtindia.nic.in/uniquepage.asp?id_pk=113 (accessed 20 December 2013).

DBT-ICMR (Department of Biotechnology and Indian Council of Medical Research) (2007) *Guidelines for Stem Cell Research and Therapy* (New Delhi: DBT-ICMR).

Dennis, C. (2002) 'China: Stem Cells Rise in the East', *Nature*, 419, 334–6.

de Vries, G. (2007) 'What is Political in Sub-politics? How Aristotle Might Help STS', *Social Studies of Science*, 37, 5, 781–809.

Dickenson, D. (2007) *Property in the Body. Feminist Perspectives* (Cambridge: Law, Medicine and Ethics).

Dikötter, F. (1998) *Imperfect Conceptions. Medical Knowledge, Birth Defects and Eugenics in China* (London: C. Hurst & Co.).

Döring, O. and Chen, R. (eds) (2002) *Advances in Chinese Medical Ethics: Chinese and International Perspectives* (Hamburg: Mitteilungen des Institutes für Asienkunde).

Doering, O. (2003) 'Chinese Researchers Promote Biomedical Regulations: What are the Motives of the Biopolitical Dawn in China and Where are they Heading?', *Kennedy Institute of Ethics Journal*, 14, 1, 39–46.

Ehara, Y. (1985) *Josei Kaihô to iu Shisô* (Tokyo: Keisôshobô).

Emery, A. E. H. (1991) 'Population Frequency of Inherited Neuromuscular Disease, a World Survey', *Neuromuscular Disorders*, 1, 19–29.

Fiester, A. (2013) 'A Dubious Export: The Moral Perils of American-style Ethics Consultation', *Bioethics*, 27, 1, ii–iii.

Foreign Press Centre (2004) 'Government Council Okays Human Embryo Cloning for Basic Research', 6 July [electronic document].

Foucault, M. (1991) [1978] 'Governmentality', in Burchell, G., Gordon, C. and Miller, P. (eds) *The Foucault Effect: Studies in Governmentality*, pp. 87–104 (Chicago, IL: University of Chicago Press).

Fox, R. and Swazey, J. (2008) *Observing Bioethics* (Oxford: Oxford University Press).

Franklin, S. (2003) 'Ethical Biocapital: New Strategies of Cell Culture' in Franklin, S. and Lock, M. (eds) *Remaking Life & Death. Toward an Anthropology of the Biosciences*, pp. 97–127 (Oxford: James Currey; Santa Fe, NM: School of American Research Press).

Franklin, S. and Roberts, C. (2006) *Born and Made. An Ethnography of Pre-implantation Genetic Diagnosis* (Princeton, NJ: Princeton University Press).

Fukuda, E. and Nakai, M. (2008) *Nihon no Yaku ha Dokoka Okashii! [There is Something Unusual About Japanese!]* (Tokyo: Seishisha).

Gangolli, L. V., Duggal, R. and Shukla, A. (eds) (2005) *Review of Healthcare in India* (Mumbai: Centre for Enquiry into Health and Allied Themes).

Gardner-Medwin, D. and Sharples, P. (1989) 'Some Studies of the Duchenne and Autosomal Recessive Types of Muscular Dystrophy', *Brain Development*, 11, 91–7.

Giddens, A. (1999) 'Risk and Responsibility', *Modern Law Review*, 62, 1–10.

Gieryn, T. F. (1983) 'Boundary-work and the Demarcation of Science From Non-science: Strains and Interests in Professional Ideologies of Scientists', *American Sociological Review*, 48, 6, 781–95.

Gieryn, T. F. (1999) *Cultural Boundaries of Science: Credibility on the Line* (Chicago, IL: Chicago University Press).

Gottweis, H., Salter, B. and Waldby, C. (2009) *The Global Politics of Human Embryonic Stem Cell Science. Regenerative Medicine in Transition* (Basingstoke; Boston, MA: Palgrave Macmillan).

Greenhalgh, S. (2008) *Just One Child: Science and Policy in Deng's China* (Berkeley, CA: University of California Press).

Guo, S. (2009) 'The "Gene War of the Century": A Cautionary Tale', in *Bionet Symposium 'Biobanking and Personal Genomics'*, Shenzhen, China, 27–29 April 2009.

Gupta, J. A. (2007) 'Private and Public Eugenics: Genetic Testing and Screening in India', *Journal of Bioethical Inquiry*, 4, 3, 217–28.

Harper, P.S. (1988) 'Genetic Counseling in Mendelian Disorders', in Harper, P. S. (ed.) *Practical Genetic Counseling*, pp. 18–41 (London: Wright).

Harris, S. (2002) 'Asian Pragmatism', *EMBO Reports*, 3, 9, pp. 816–17.

Harris, J. (2011) *Enhancing Evolution. The Ethical Case for Making Better People* (Princeton, NJ: Princeton University Press).

Hayden, C. (2003) *When Nature Goes Public: The Making and Unmaking of Bioprospecting in Mexico* (Princeton, NJ: Princeton University Press).

Hayden, C. (2007) 'Taking as Giving Bioscience, Exchange, and the Politics of Benefit-sharing', *Social Studies of Science*, 37, 5, 729–58.

Helmreich, S. (2007) 'Blue-green Capital, Biotechnological Circulation and an Oceanic Imaginary: A Critique of Biopolitical Economy', *BioSocieties*, 2, 287–302.

Helmreich, S. (2008) 'Species of Biocapital', *Science as Culture*, 17, 4, 463–78.

Henneman, L., Bramsden, I., Van Os, T. A. M., et al. (2001) 'Attitudes Towards Reproductive Issues and Carrier Testing Among Adult Patients and Parents of Children with Cystic Fibrosis (CF)', *Prenatal Diagnosis*, 21, 1–9.

Hennig, W. (2009) 'Research in China. Experiences from 23 Years of Molecular Genetics Research in Shanghai', *EMBO Reports*, 10, 6, 545–50.

Hesketh, T. and Xing, Z.-W. (2000) 'Human Population Growth: China's One Child Family Policy is Changing', *British Medical Journal*, 320, 7232, 443.

Hill, R. H. (2007) 'The Emergence of Laboratory Safety', *Journal of Chemical Health & Safety*, 14, 3, 14–19.

Hishiyama, Y. (2003) *Seimei Ronri Handobukku (Handbook of Bioethics)* (Tokyo: Tsukiji-Shokan).

Holland, S., Lebacqz, K. and Zoloth, L. (2001) *The Human Embryonic Stem Cell Debate* (Cambridge, MA; London: MIT Press).

Horlick-Jones, T. (2007) 'On the Signature of New Technologies: Sociality, Materiality and Practical Reasoning', in Flynn, R. and Bellaby, P. (eds) *Risk and the Public Acceptance of New Technologies*, pp. 41–65 (Basingstoke: Palgrave Macmillan).

HUGO (Human Genome Organization) Committee (1995) *Human Genome Diversity (HGD) Project: Summary Document, Report of the International Planning Workshop of the HGD Project*, Porto Conto, Sardinia, Italy, 9–12 September 1993 (London: HUGO Europe).

ICMR (Indian Council of Medical Research) (2000) 'Ethical Guidelines for Biomedical Research on Human Subjects', *ICMR Bulletin*, 30, 10, 107–16.

Ida, R. (2002) 'Ethical Questions of the Human Embryonic Stem Cells Research', *Rinsho Shinkeigaku [Clinical Neurology]*, 42, 11, 1147–8.

ISCSI (International Stem Cell Summit India) (2008) 'Proceedings of the 1st ISCSI', Verma, R. S. (ed.), Tamil Nadu, India, 16–18 November 2008.

Ivry, T. (2006) 'At the Back Stage of Prenatal Care: Japanese Ob-gyns Negotiating Prenatal Diagnosis', *Medical Anthropology Quarterly*, 20, 4, 441–68.

Japan Times Weekly 'A Bill for Integrated Welfare', available at: http://weekly.japantimes.co.jp/ed/a-bill-for-integrated-welfare (accessed 28 May 2005).

Jasanoff, S. (2005) *Designs on Nature: Science and Democracy in Europe and the United States* (Princeton, NJ: Princeton University Press).

Jayaram, K. S. (2005) 'Biotech Boom', *Nature Biotechnology*, 23, 9, 1183–4.

Jones, G., Straughan, P. and Chan, A. (2009) *Ultra-low Fertility in Pacific Asia. Tends, Causes and Policy Issues* (Abingdon; New York: Routledge).

JSTA (Japan Science and Technology Agency) (2009) 'Strategic Sector: Creating Fundamental Technologies for Advanced Medicine Through Generation and Regulation of Stem Cells, Based on Cellular Reprogramming', available at: http://www.jst.go.jp/kisoken/teian/en/mokuhyo/h20-ips.html (accessed 3 December 2013).

Kato, K. (2005) 'The Ethical and Political Discussions on Stem Cell Research in Japan', in Bender, W., Hauskeller, C and Manzei, A. (eds) *Grenzüberschreitungen [Crossing Borders]*, pp. 369–82 (Münster: Agenda Verlag).

Kato (2007) 'Silence Between Patients and Doctors: The Issue of Self-determination and Amniocentesis in Japan', *Genomics, Society and Policy*, 3, 3, 28–42.

Kato, M. (2009a) 'Quality of Offspring? Socio-cultural Factors, Pre-natal Testing and Reproductive Decision-making in Japan', *Culture, Health & Sexuality*, 2010, 12, 2, 177–9.

Kato, M. (2009b) 'Culture of Marriage, Reproduction and Genetic Testing in Japan', *Biosocieties*, 4, 2–3, 115–27.

Kato, M. and Sleeboom-Faulkner, M. (2011) 'Meanings of the Embryo in Japan: Narratives of IVF Experience and Embryo Ownership', *Issues of Social Health & Illness*, 33, 3.

Kayukawa, J. (2003) *Kuroun Ningen [Human Cloning]* (Tokyo: Koubunsha Shinsho).

KBA (Korean Bioethics Association) (2004) 'Medical and Biotechnological Research Must Accord to Bioethical Principles', *Statement Special Ethics Committee*, 22 May 2004.

Kevles, D. J. (1985) *In the Name of Eugenics: Genetics and the Uses of Human Heredity* (New York: Knopf).

Khan, A. (2007) 'Are we Stemming Growth?', available at: http://pharma.financialexpress.com/20070615/management02.shtml (accessed 3 December 2013).

Kiatpongsan, S. and Sipp, D. (2008) 'Offshore Stem Cell Treatments', available at: http://www.nature.com/stemcells/2008/0812/081203/full/stemcells.2008.151.html (accessed 13 January 2009).

Kiefer, C. W. (1987) 'Care of the Aged in Japan', in Norbeck, E. and Lock, M. (eds) *Health, Illness, and Medical Care in Japan: Cultural and Social Dimensions*, pp. 89–109 (Honolulu, HI: University of Hawaii Press).

Kim, H. (2004) 'A Study on the Policy Network of Bioethics Agenda Setting in Korea', *Journal of Korean Association of Bioethics*, 4, 1, 1–20 [in Korean].

Kim, T. H. (2008a) 'How Could a Scientist Become a National Celebrity? Nationalism and Hwang Woo-Suk Scandal', *East Asian Science, Technology and Society: an International Journal*, 2, 27–45.

Kim, L. (2008b) 'Governing Discourses of Stem Cell Research in the UK and South Korea: 1997–2009', in *4th BIONET Conference*, Changsha, China, 1–3 April, 2008.

Kipnis, A. (2007) 'Neoliberalism Reified: *Suzhi* Discourse and Tropes of Neoliberalism in the People's Republic of China', *Journal of the Royal Anthropological Institute*, 13, 383–400.

Kitagawa, F. and Woolgar, I. (2008) 'Regionalisation of Innovation Policies and New University–industry Links in Japan: Policy Review and New Trends', *Prometheus*, 26, 1, 55–67.

Knoppers, B. M. (2003) *Populations and Genetics: Legal and Socio-Ethical Perspectives* (Leiden; Boston, MA: Martinus Nijhoff Publishers).

Kohrman, M. (1999) 'Grooming Quezi: Marriage Exclusion and Identity Formation Among Disabled Men in Contemporary China', *American Ethnologist*, 26, 4, 890–909.

Kohrman, M. (2005) *Bodies of Difference. Experiences of Disability and Institutional Advocacy in the Making of Modern China* (Berkeley, CA: California University Press).

Kôseishô imukyoku (1955) *80 Years' History of Medicine* (Tokyo: Kôseishô Imukyoku (Bureau of Medical Affairs)).

Kulkarni, N. (2008) 'Asia to Dominate Adult Stem Cell Commercialization', available at: http://cybermediaservicesdigitalmag.com/admin/magazines/vol3_issue19/BSA_pages/p14.pdf (accessed on 13 May 2011).

Kuo, W. H. (2008) 'Understanding Race at the Frontier of Pharmaceutical Regulation: An Analysis of the Racial Difference Debate at the ICH', *Journal of Law, Medicine and Ethics* 36, 3, 498–505.

Lafleur, W. (1992) *Liquid Life. Abortion and Buddhism in Japan* (Princeton, NJ: Princeton University Press).

Lander, B., Thorsteinsdottir, H., Singer, P. A. and Daar, A. S. (2008) 'Harnessing Stems Cells for Health Needs in India', *Cell Stem Cells*, 3, 1, 11–15.

Latimer, J. (2007) 'Diagnosis, Dysmorphology, and the Family: Knowledge, Motility, Choice', *Medical Anthropology*, 26, 97–138.

Latour, B. (2005) *Reassembling the Social. An Introduction to Actor-Network-Theory* (Oxford: Oxford University Press).

Lee, S. (2003) 'Racial Profiling of DNA Samples: Will it Affect Scientific Knowledge About Human Genetic Variation?', in Knoppers, M. (ed.) *Populations and Genetics: Legal and Socio-Ethical Perspectives*, pp. 231–44 (Leiden; Boston, MA: Martinus Nijhoff Publishers).

Lei, Y. (2006) *Yunqian-hou yi yu ji* [*Do's and Don'ts Before and After Pregnancy*] (Beijing: Zhong-Yi Guji Chubanshe).

Li, J. (2006) 'Free Testing for Thalassemia in Nanning', *Zhongguo Renkou* [China's Population], available at http://www.chinapop.gov.cn/ (accessed 4 December 2013).

Li, C. and Zhou, J. (2005) 'Hubei Chanqian Yichuan Jiance Yaoqiu Zhunruzhi. ['Hubei Asks for Permission for Prenatal Genetic Testing'], Xinhua News, 9 Dec.

Li, X. and Wang, L. (2006) *Huaiyun Zhunbei Yiji: Zhuanjia Zhonggao [Do's and Don'ts for Pregnancy Preparation: Expert Advice]* (Beijing: Huaxue Gongye Chubanshe).

Liu, S. (2005) 'Zhongguo de Xingbie Pianhao' ['Gender preference of Chinese couples'], *Renkou Yanjiu [Population Research]*, 3, 1–11.

Lock, M. (1980) *East Asian Medicine in Urban Japan: Varieties of Medical Experience* (Berkeley, CA: University of California Press).

Lock, M. (1994) 'Interrogating the Human Diversity Genome Project', *Social Science & Medicine*, 39, 603–6.

Lock, M. (1998) 'Perfecting Society: Reproductive Technologies, Genetic Testing, and the Planned Family in Japan' in Lock, M. and Kaufert, P. (eds) *Pragmatic Women and Body Politics*, pp. 206–39 (Cambridge: Cambridge University Press).

Lock, M. (2002) *Twice Dead. Organ Transplants and the Reinvention of Death* (Berkeley, CA; London: University of California Press).

Lock, M. (2003) 'Utopias of Health Eugenics, and Germline Engineering' in Nichter, M. and Lock, M. (eds) *New Horizons in Medical Anthropology* pp. 240–66 (London: Routledge).

Lock, M. and Kaufert, P. (1998) *Introduction. Pragmatic Women and Body Politics* (Cambridge: Cambridge University Press).

Long, S. O. (2005) *Final Days: Japanese Culture and Choice at the End of Life* (Honolulu, HI: University of Hawaii Press).

Macer, D. R. (1997) 'Bioethics and Genetic Diversity from the Perspective of UNESCO and Non-governmental Organizations', in Knoppers, B. M., Laberge, C. and Hirtle, M. (eds) *Human DNA: Law and Policy. International and Comparative Perspectives*, pp. 265–74 (The Hague: Kluwer Law International).

Majumder, P. P. (2000) 'Genes, Diversities and Peoples of India', in Macer, D. (ed.) *Ethical Challenges As we Approach the End of the Human Genome Project*, pp. 20–33 (Tsukuba: Eubios Ethics Institute).

Mandot, S., Khurana, V. L. and Sonesh, J. K. (2009) 'Sickle Cell Anaemia in Garasia Tribals of Rajasthan', *Indian Pediatrics*, 46, 239–40.

Mao, X. (1998) 'Chinese Geneticists' View of Ethical Issues in Genetic testing and Screening: Evidence for Eugenics in China', *American Journal of Human Genetics*, 63, 688–95.

Mao, X. and Wertz, D. C. (1997) 'China's Genetics Services Provider's Attitudes Towards Several Ethical Issues: A Cross-cultural Survey', *Clinical Genetics*, 52, 100–9.

Marshall, E. (1997) 'Gene Prospecting in Remote Populations', *Science*, 278, 5338, 565.

Marteau, T. M. and Johnston, M. (1986) 'Determinants of Beliefs About Illness: A Study of Parents of Children with Diabetes, Asthma, Epilepsy, and no Chronic Illness', *Journal of Psychosomatic Research*, 30, 673–83.

Marzuki, S., Sudoyo, H., Suryadi, H., Setianingsih, I. and Pramoonjago, P. (2003) 'Human Genome Diversity and Disease on the Island Southeast Asia' in Marzuki, S., Verhoef , J. and Sippe, H. (eds) *Tropical Diseases. From Molecular to Bedside* pp. 1–19 (New York: Kluwer Academic/Plenum Publishers).

Masui, T. (2009) 'Trust and the Creation of Biobanks: Biobanking in Japan and the UK', in Sleeboom-Faulkner, M. (ed.) *Human Genetic Biobanking in Asia* pp. 66–91 (London; New York: Routledge).

MEXT (Ministry of Education, Culture, Sport, Science and Technology), Council for Science and Technology, Bioethics Committee (2000) 'Fundamental principles of Research on the Human Genome', available at: http://www.mext. go.jp/a_menu/shinkou/shisaku/fundamen.htm (accessed April 2008).

MEXT (Ministry of Education, Culture, Sport, Science and Technology) (2008) 'Regulation for Therapeutic Cloning for Research Purposes', available at: http://www.lifescience.mext.go.jp/files/pdf/2_32.pdf (accessed 4 December 2013).

MEXT (Ministry of Education, Culture, Sport, Science and Technology) (2009) 'Guidelines for the Establishment and Utilization of Human Embryonic Stem Cell Lines', available at: http://www.lifescience.mext.go.jp/files/pdf/n743_01. pdf (accessed 20 December 2013).

MoH (Ministry of Health) (2003a) *Measures for the Administration of Prenatal Diagnosis Technology, Article 24* (Beijing: MoH).

MoH (Ministry of Health) (2003b) 'Guidelines for Genetic Counselling', available at: http://www.healthychildren.org.cn/actionplan/fagui/fagui.htm (accessed 10 October 2011).

MoH (Ministry of Health) (2003c) 'Ethical Guiding Principles on Human Embryonic Stem Cell Research (2003-460)' (Promulgated by the Ministry of Science and Technology and the Ministry of Health, People's Republic of China on December 24, 2003), available at: http://www.chinaphs.org/bioethics/regulations_&_laws.htm#_Toc113106142 (accessed 10 October 2011).

MoHWL (Ministry of Health, Labour and Welfare) (2005) 'Kekkon to Shussan ni Kansuru Zenkoku Choosa' ['National Survey on Marriage and Giving Birth'], available at: http://www.mhlw.go.jp/shingi/2006/09/dl/s0929-6b2.pdf (accessed on 21 November 2013).

MoHWL (Ministry of Health, Welfare and Labour) (2013), available at: http://www.mhlw.go.jp/ (accessed 20 December 2013).

Morioka, M. (1995) 'Bioethics and Japanese Culture: Brain Death, Patients' Rights, and Cultural Factors', *Eubios Journal of Asian and International Bioethics*, 5, 87–91.

Mudur, G. (1996) 'India Concerned at Export of Genetic Material', *British Medical Journal*, 312, 464.

Nakatsuji, N. (2007) 'Irrational Japanese Regulations Hinder Human Embryonic Stem Cell Research', available at: http://www.nature.com/stemcells/2007/0708/070809/full/stemcells.2007.66.html (accessed 4 December 2013).

Nature Editorial (2013) 'Unknown Territory: Japan is Making an Overdue Effort to Regulate Experimental Stem-cell Treatments. A Clearly Defined Legal Framework is Needed to Protect Patients', *Nature*, 494, 5.

New York Times (2010) 'Will China Achieve Science Supremacy?', available at: http://roomfordebate.blogs.nytimes.com/2010/01/18/will-china-achieve-science-supremacy/ (accessed 4 December 2013).

Nie, J.-B. (2005) *Behind the Silence. Chinese Voices on Abortion* (Lanham, MD; Oxford: Rowman & Littlefield).

Nishikawa, S.-I., Goldstein, R. and Nierras, C. R. (2008) 'The Promise of Human Induced Pluripotent Stem Cells for Research and Therapy', *Nature Reviews Molecular Cell Biology*, 9, 725–9.

Nomaguchi, C., Hasegawa, T., Yamaguchi, M. and Itai, K. (2007) 'Genetic Counselling of a Client with MEN [Multiple Endocrine Neoplasia Type 1] and his Family', *Japanese Journal of Genetic Counselling*, 57.

Norgren, T. (2001) *Abortion Before Birth Control. The Politics of Reproduction in Postwar Japan* (Princeton, NJ: Princeton University Press).

Numabe, H. (2007) 'Iden Counselling to Tabunka' ['Genetic Counseling and Chance'], in Mizutani, S., Yoshida, M. and Ozasa, Y. (eds) *Iden Shinryô o Torimaku Shakai: Sono Kagakuteki/Rinriteki Approach*, pp. 103–12 [*The Society That Struggles With Genetic Therapy: A Scientific/Ethical Approach*] (Tokyo: Buren Shuppan).

Ong, A. and Collier S. (2005) 'Introduction', in Collier, S. and Ong, A. (eds) *Global Assemblages: Technology, Politics and Ethics and Anthropological Problems* pp. 3–21 (Malden, MA: Blackwell).

Ong, A. and Chen, N. (2010) 'Introduction to Asian Biotech Bioeconomies & Communities of Fate', in Ong, A. and Chen, N. (eds) *Asian Biotech. Bioeconomies & Communities of Fate*, pp 1–51 (Durham, NC: Duke University Press).

Ozasa, Y. (2006) 'Yôsuikensa, Iryô Genba Deno Shien. Post-genomic Jidai ni Okeru Seibutsu-igaku to Gender Ikagaku Gijutsu ni Okeru Literacy o Kangaeru' ['Amniocentesis Support at Medical Locations. A Study on Literacy in Life Science and Gender Aspects of Medical Technology in the Post-genomic Era'], in Senba, Y. and Muto, K. (eds) *Ochanomizu University the 21st Century COE Programme*, pp. 79–85 (Tokyo: Frontiers of Genomic Studies).

Palmer, L. J. (2007) 'UK Biobank: Bank on it', *The Lancet*, 369, 9578, 1980–2.

Pandya, S. (2008) 'Stem Cell Transplantation in India: Tall Claims, Questionable Ethics', *Indian Journal of Medical Ethics*, 1, 15–17.

Park, P. K. (2006) 'A Study on the Historical Development of Research Community in Korea: Focused on the Government Supported Institutes', *Science and Technology Studies*, 6, 1, 119–51.

Passier, R., van Laake, L. W. and Mummery, C. L. (2008) 'Stem Cell Based Therapy and Lessons From the Heart', *Nature*, 453, 322–9.

Patra, P. K. and Sleeboom-Faulkner, M. (2007) 'Genetic Biobanking in India – A Community Based Perspective on Ways and Means of Data Generation', *Taiwan Journal of Law and Technology Policy*, 4, 1, 67–97.

Patra, P. K. and Sleeboom-Faulkner, M. (2009a), 'The Indian Genomic Biobank Initiative and Emerging Bioethical Issues: A Community Based Perspective', in Sleeboom-Faulkner, M. (ed.) *Human Genetic Biobanks in Asia: Politics of Trust and Scientific Advancement* pp. 151–67 (London; New York: Kegan Paul).

Patra, P. K. and Sleeboom-Faulkner, M. (2009b) 'Bionetworking: Experimental Stem Cell Therapy and Patient Recruitment in India', *Anthropology & Medicine*, 16, 2, 147–63.

Patra, P. K. and Sleeboom-Faulkner, M. (2010) 'Population Genetic Screening for Sickle Cell Anaemia Among Rural and Tribal Communities in India: The Limitations of Socio-ethical Choice', in Sleeboom-Faulkner, M. (ed.) *Frameworks of Choice: Predictive & Genetic Testing in Asia*, pp. 65–90 (Amsterdam: University of Amsterdam Press).

Patra, P. K. and Sleeboom-Faulkner, M. (2011) 'Recruiter-patients as Ambiguous Symbols: Bionetworking and Stem Cell Therapy in India', *New Genetics and Society*, 30, 2, 155–66.

Partridge, W. M. (2003) 'Translational Science: What it is and Why it is so Important', *Drug Discovery Today,* 8, 18, 813–15.

Peterson, A. L. (2001) *Being Human. Ethics, Environment, and Our Place in the World* (Berkeley, CA; London: California University Press).

Petryna, A. (2005) 'Ethical Variability: Drug Development and Globalizing Clinical Trials', *American Ethnologist*, 32, 2, 183–97.

Petryna, A. (2007) 'Globalizing Human Subject Research', in Petryna, A., Lakoff. A. and Kleinman, A. (eds) *Global Pharmaceuticals – Ethics, Markets, Practices* pp. 33–60 (Durham, NC: Duke University Press).

Petryna, A. (2009) *When Experiments Travel* (Princeton, NJ: Princeton University Press).

Pilnick, A. (2002) *Genetics and Society: An Introduction* (Buckingham; Philadelphia, PA: Open University Press).

Pomfret, J. and Nelson, D. (2000) 'An Isolated Region's Genetic Mother Lode', *Washington Post*, 20 Dec.

Porter, G. (2010) 'Genetic Tests and Insurance in Japan: The Case for a Regulatory Framework', in Sleeboom-Faulkner, M. (ed.) *Frameworks of Choice: Predictive & Genetic Testing in Asia* pp. 145–66 (Amsterdam: University of Amsterdam).

Prainsack, B. and Siegal, G. (2006) 'The Rise of Genetic Couplehood? A Comparative View of Premarital Genetic Testing', *Biosocieties*, 1, 17–36.

Pun, N. (2003) 'Subsumption or Consumption? The Phantom of Consumer Revolution in "Globalizing" China', *Cultural Anthropology*, 18, 469–92.

Qin, Z. (2004) *Jiankang Huaiyun Mei yi Tian* [*Healthy Pregnancy Every Day*] (Beijing: Zhonghua Guji Chubanshe).

Qiu, R. (1987) *Shengming Lunlixue* (*Bioethics*) (Shanghai: Shanghai Renmin Chubanshe).

Qiu, R. Z. (ed.) (2006) *Bioethics: Asian Perspectives: A Quest for Moral Diversity* (Dordrecht: Kluwer).

Rangarajan, R. (2006) 'Hospitals and Clinics – Lifeline Acquires 15 Clinics in TN (Tamil Nadu)', available at: http://archives.chennaionline.com/health/Hospital sandClinics/2006/03lifeline.asp (accessed 14 January 2009).

Robertson, J. (2005) 'Dehistoricizing History. The Ethical Dilemma of "East Asian Bioethics"', *Critical Asian Studies*, 37, 2, 233–50.

Rogaski, R. (2010) 'Vampires in Plague Land: The Multiple Meanings of *Weisheng* in Manchuria', in Leung, A. K. C. and Furth, C. (eds) *Health and Hygiene in Chinese East Asia. Policies and Publics in the Long Twentieth Century*, pp. 132–58 (Durham, NC; London: Duke University Press).

Rose, N. (2007) *The Politics of Life Itself. Biomedicine, Power, and Subjectivity in the Twenty-First Century* (Princeton, NJ: Princeton University Press).

Rose, N. and Novas, C. (2005) 'Biological citizenship', in Ong, A. and Collier, S. (eds) *Global Assemblages: Technology, Politics and Ethics as Anthropological Problems* pp. 439–63 (Malden, MA: Blackwell Publishing).

Royal Society (2011) 'Knowledge, Networks and Nations', report by the Royal Society, available at: http://royalsociety.org/uploadedFiles/Royal_Society_ Content/Influencing_Policy/Reports/2011-03-28-Knowledge-networks-nations.pdf (accessed 5 December 2013).

Sabarini, P. (2010) 'Sangkot Marzuki: The Brilliant Professor Behind Eijkman', *The Jakarta Post*, available at:, http://www.thejakartapost.com/news/2010/02/04/ sangkot-marzuki-the-brilliant-professor-behind-eijkman.html (accessed 4 December 2013).

Saegusa, A. (1999) 'Looser Ties Urged for Japanese Scientists', *Nature*, 398, 643, 6729.

Salter, B. (2008) 'Governing Stem Cell Science in China and India: Emerging Economies and the Global Politics of Innovation', *New Genetics and Society*, 27, 2, 145–59.

Salter, B. and Salter, C. (2007) 'Bioethics and the Global Moral Economy: The Cultural Politics of Human Embryonic Stem Cell Science', *Science, Technology & Human Values*, 32, 554, 554–81.

Salter, B., Cooper, M., Dickins, A. and Cardo, V. (2007) 'Stem Cell Science in India: Emerging Economies and the Politics of Globalization', *Regenerative Medicine*, 2, 1, 75–89.

Sándor, J. (2003) *Society and Genetic Information. Codes and Laws in the Genetic Era* (Budapest; New York: Central European University Press; CPS Books).

Sangren, P. S. (1995) '"Power" Against Ideology: A Critique of Foucaultian Usage', *Current Anthropology* 10, 1, 3–40.

Sato, R. and Iwasawa, M. (2006) 'Contraceptive Use and Induced Abortion in Japan. How is it so Unique Among the Developed Countries?', *The Japanese Journal of Population*, 4, 1, 33–54.

Savulescu, J. (2009) 'The Human Prejudice and the Moral Status of Enhanced Beings: What Do we Owe the Gods?', in Savulescu, J. and Bostrom, N. (eds) *Human Enhancement*, pp. 211–49 (Oxford: Oxford University Press).

Scharping, T. (2003) *Birth Control in China 1949–2000: Population Policy and Demographic Development* (London: Routlege/Curzon).

Sharma, A. (2006) 'Stem Cell Research in India: Emerging Scenario and Policy Concerns', *Asian Biotechnology and Development Review*, 8, 3, 43–53.

Shimazono, S. (2008) 'Why Must we be Prudent in Research Using Human Embryos: Differing Views of Human Dignity', in LaFleur, W. R., Boehme, G. and Shimazono, S. (eds) *Dark Medicine*, pp. 201–22 (Bloomington, IN: Indiana University Press).

Shineha, R., Kawakami, M., Kawakami, K., Nagata, M., Tada, T. and Kato, K. (2010) 'Familiarity and Prudence of the Japanese Public with Research into Induced Pluripotent Stem Cells, and Their Desire for its Proper Regulation', *Stem Cell Reviews and Reports*, 6, 1–7.

Shroff, G. (2005) *Human Embryonic Stem Cells – A Revolution in Therapeutics* (New Delhi: Nu-Tech Mediworld).

Simpson, B. (2010) 'A "Therapeutic gap": Anthropological Perspectives on Prenatal Diagnostics and Termination in Sri Lanka', in Sleeboom-Faulkner, M. (ed.) *Frameworks of Choice: Predictive & Genetic Testing in Asia* pp. 27–42 (Amsterdam: University of Amsterdam).

Sivin, N. (1980) 'The Theoretical Background of Laboratory Alchemy', in Needham, J., Ho, D.-D., Lu, G.-D. and Sivin, N.. (eds) *Science and Civilisation in China*, vol. V, part 5, pp. 210–305 (Cambridge: Cambridge University Press).

Slamet, L. S. and Elengold, M. A. (2002) 'Regulating Biotechnology Products', available at: http://www.docstoc.com/docs/45702611/Regulating-biotechnology-products (accessed 20 December 2013).

Sleeboom, M. (2004) *Genomics in Asia: A Clash of Bioethical Interests?* (London: Kegan Paul).

Sleeboom, M. (2005) 'The Harvard Case of Xu Xiping: Exploitation of the People, Scientific Advance or Genetic Theft?', *New Genetics and Society*, 20, 57–87.

Sleeboom-Faulkner, M. (2007) 'Collecting Families: An Institutional Approach to Human Genetic Biobanking in Indonesia', *The Asian Social Science Journal*, 36, 1, 626–56.

Sleeboom-Faulkner, M. (2008a) 'The Changing Nature of Ideology in the Life Sciences in the People's Republic of China (PRC). Case Studies of Human Cloning and Human Embryonic Stem Cell Research (hESR) in Medical Textbooks (1996–2005)', *Biosocieties*, 3, 21–36.

Sleeboom-Faulkner, M. (2008b) 'Claiming the Futures of Human Embryonic Stem Cell Research in Japan: Minority Voices and Their Amplifiers', *Science as Culture*, 17, 1, 95–7.

Sleeboom-Faulkner, M. (2010a) 'Eugenic Birth and Foetal Education: How Instructions for Premarital Testing in Mainland China Miss Their Target Among Rural Households', *The China Journal*, 64, 121–41.

Sleeboom-Faulkner, M. (2010b) 'National Risk Signatures and Human Embryonic Stem Cell Research in Mainland China', *Health, Risk and Society*, 12, 5 491–511.

Sleeboom-Faulkner, M. (2010c) 'Boundary-making Between Science, Society and the World. The Case of Stem Cell Research in China', *East Asia Journal for Science and Technology Studies (EASTS)*, 4, 1, 31–51.

Sleeboom-Faulkner, M. (2010d) 'Contested Embryonic Culture in Japan – Public Discussion, and Human Embryonic Stem Cell Research in an Aging Welfare Society', *Medical Anthropology*, 29, 1.

Sleeboom-Faulkner, M. (2010e) 'Frameworks of Choice: The Ramification of Predictive and Genetic Testing in Asia', in Sleeboom-Faulkner, M. (ed.) *Frameworks of Choice: Predictive & Genetic Testing in Asia*, pp. 11–26 (Amsterdam: University of Amsterdam).

Sleeboom-Faulkner, M. (2011a) 'Genetic Testing, Governance, and the Family in the PRC', *Social Science and Medicine*, 72, 1802–9.

Sleeboom-Faulkner, M. (2011b) 'Regulating Cell Lives in Japan: Avoiding Scandal and Sticking to Nature', *New Genetics and Society*, 30, 3, 227–40.

Sleeboom-Faulkner, M. (2013a) 'Latent Science Collaboration: Strategic Cultures of Bioethical Capacity Building in Mainland China's Stem Cell World', *Biosocieties*, 8, 7–24.

Sleeboom-Faulkner, M. (2013b) 'Competitive Adaptation: Biobanking and bio-ethical governance in China Medical City (CMC)', *East Asia Journal for Science and Technology Studies*, 7,1, 125–43 .

Sleeboom-Faulkner, M. and Patra, P. K. (2008) 'The Bioethical Vacuum: National Policies on Human Embryonic Stem Cell Research in India and China', *Journal of International Biotechnology Law*, 5, 6, 221–34.

Sleeboom-Faulkner, M. and Patra, P. K. (2011) 'Experimental Stem Cell Therapy: Biohierarchies and Bionetworking in Japan and India', *Social Studies of Science*, 41, 5, 645–66.

Sleeboom-Faulkner, M. and Hwang, S. (2012) 'Governance of Stem Cell Research: Public Participation and Decision-making in China, Japan, South-Korea, Taiwan and the UK', *Social Studies of Science*, 42, 5, 684–708.

Slingby, B.T., Nagao, N. and Akabayashi, A. (2004), 'Administrative Legislation in Japan: Guidelines on Scientific and Ethical Standards', *Cambridge Quarterly of Healthcare Ethics*, 13, 245–53.

Som, N. (2007) 'Chennai: The Haven of Medical Research', available at: http://www.expresspharmaonline.com/20070615/healthcare01.shtml (accessed 13 January 2009).

Song, P. (2011) 'The Proliferation of Stem Cell Therapies in Post-Mao China: Problematizing Ethical Regulation', *New Genetics and Society*, 30, 2, 141–53.

Srinivasan, S. (2004) 'Indian Guinea Pigs for Sale: Outsourcing Clinical Trials', available at: http://www.indiaresource.org/issues/globalization/2004/indian guineapigs.html (accessed 20 December 2013).

Stapel, J. (ed.) (2003) *Scientific Programme Indonesia–Netherlands (SPIN)* (Amsterdam: KNAW).

Stemerding, D. and Nelis, A. (2006) 'Cancer Genetics and its "Different Faces of Autonomy"', *New Genetics and Society*, 25, 1, 1–19.

Stock, G. (2003) *Redesigning Humans. Choosing our Children's Genes* (London: Profile Books).

Strauer, B. E. and Kornowski, R. (2003) 'Stem Cell Therapy in Perspective', *Journal of the American Heart Association*, 107, 929–934.

Sui, S. (2007) *'Guanyu Zhongguo Xingbiebi Shengheng de Sikao'* ['Reflection on Balancing the Sex Ratio in China'], *Zhongguo Yixue Lunlixue* [*Chinese Medical Ethics*], 2, 89–91.

Sui, S. and Sleeboom-Faulkner, M. (2007) 'Commercial Genetic Testing in Mainland China: Social, Financial and Ethical Issues', *Journal of Bioethical Inquiry*, 4, 3, 229–37.

Sui, S. and Sleeboom-Faulkner, M. (2010a) 'Choosing Offspring – Case Studies of Prenatal Genetic Testing for Thalassemia and the Reproduction of a "Saviour Sibling" in China', *Culture, Health and Sexuality*, 12, 2, 167–75.

Sui, S. and Sleeboom-Faulkner, M. (2010b) 'Genetic Testing on Duchenne Muscular Dystrophy in China: Vulnerabilities of Chinese Families', in Sleeboom-Faulkner, M. (ed.) *Frameworks of Choice: Predictive & Genetic testing in Asia* pp. 167–82 (Amsterdam: University of Amsterdam).

Sunder Rajan, K. (2006) *Biocapital – the Constitution of Postgenomic Life* (Durham, NC: Duke University Press).

Sunder Rajan, K. (2007) 'Experimental Values: Indian Clinical Trials and Surplus Health', *New Left Review*, 45, 67–88.

Sustrop, M. (ed.) (2004) Special issue: 'Human Genetic Databases: Ethical, Legal and Social Issues', *Journal of the Humanities and Social Sciences*, 8, 1/2.

Tamura, C. (2007) 'Iden Counsellor: Gan wa Iden Suru ka – Kanja no Fuan ni Saisentankagaku ga Ataeru Shishin' ['Genetic Counsellors: Is Cancer Genetically Inherited?: Support for Anxious Patients Provided by the New Medical Sciences'], in Fukuhara, M. (ed.) *Gan tô Byô to Co-medical: Iryô Saizensen Kara no Teigen* [*Struggles with Cancer and Medical Treatment: Proposals from the Medical Frontline*], pp. 71–87 (Tokyo: Kôdansha).

Tang, Z. X. (2006) *Qiuzi Zhuyun Bidu* [*Child Wish, Pregnancy Aid: Necessary Reading*] (Zhuanmei: Jituan Jiangsu Kexue Jishu Chubanshe).

Tateiwa, S. (1997) *Shiteki Shoyûron* (Tokyo: Keisô Shobô).

Taussig, K., Rapp, R. and Heath, D. (2003) 'Flexible Eugenics: Technologies of the Self in the Age of Genetics', in Goodman, A., Heath, E. and Lindee, S. (eds) *Genetic Nature/Culture*, pp. 58–76 (Berkeley, CA: University of California Press).

Templeton, A. (2003) 'Human Races in the Context of Recent Human Evolution. A Molecular Genetic Perspective', in Goodman, A. H., Heath, D. and Lindee, M. S. (eds) *Genetic Nature/Culture. Anthropology and Science Beyond the Two-culture Divide*, pp. 234–57 (Berkeley, CA; London: University of California Press).

The Hindu (2006) 'A National Repository of Human Genome Research and Data Will be Set up Shortly', available at: http://www.hindu.com/2006/02/23/stories/2006022308130400.htm (accessed 5 August 2011).

The Hindustan Times (2006) 'Stem Cell Trials to Start at AIIMS', available at: http://stemcell.taragana.net/archive/stem-cell-researchin-heart-disease-and-other-ailments-at-all-india-institute-of-medical-sciences/ (accessed 21 January 2009).

Toland, P. (2007) 'China Plans "Largest Gene Bank"', available at: http://news.bbc.co.uk/1/hi/sci/tech/7046586.stm (accessed 5 December 2013).

Tomiwa, K. (1998) 'Chi'iki Iden Sôdan Sentâ no Kikaku to Kinô ni Kan Suru Kenkyû' ['A study on the Planning and Functions of Local Genetic Centers]', in Aoki, K. (ed.) *Iden sô Dan ni Kansuru Kenkyû: Kôseishô Shinshin Shôgai Kenkyû* [*A Study on Genetic Counselling: Research by the Ministry of Health, Welfare and Labour on People with Physical and Mental Disablements*], pp. 49–64 (Tokyo: Ministry of Health,Welfare and Labour).

Traphagan, J. W. (1998) 'Localizing Senility: Illness and Agency Among Older Japanese', *Journal of Cross-cultural Gerontology*, 13, 81–98.

Tri-Council of Canada (1997) *Medical Research Council, Natural Sciences and Engineering Research Council: Social Sciences and Humanities Research Council, Code of Conduct for Research Involving Humans* (Ottawa: Ministry of Supply and Services Canada).

Tsuge, A. (2008) 'Can iPS Cells Break Through Contradictions in Japan?', in *The Social Regulation of Stem Cell Research: Looking Beyond Regulatory Exteriors in Asia*, Falmer, UK, 15 December 2008.

Tsuge, A. (2010) 'How Japanese Women Narrate Their Experiences of Prenatal Testing: Ultrasound, Maternal Serum Screening, and Amniocentesis', in Sleeboom-Faulkner, M. (ed.) *Frameworks of Choice: Prenatal and Genetic Testing in Asia*, pp. 109–24 (Amsterdam: University of Amsterdam Press).

Tsuji, K. (2003) 'Women's Experiences of Subsequent Pregnancy and Childbirth Following Delivery of a Child with Down's Syndrome', *Journal of the Japanese Academy of Nursing Science*, 23, 1, 46–56.

Tsukamoto, Y. (2005) *Iryô no Naka no Ishi Kettei: Shusseizenshindan* (Tokyo: Kôchishobô).

UNDP-Human Development Report (2001) 'Making New Technologies Work for Human Development', available at: http://hdr.undp.org/reports/global/2001/en/ (accessed 5 December 2013).

Unschuld, P. (1992) 'Epistemological Issues and Changing Legitimation: Traditional Chinese Medicine in the Twentieth Century', in Leslie, C. and Young, A. (eds) *Paths to Asian Medical Knowledge* (Berkeley, CA; Oxford: University of California Press).

Urtizberea, J. A., Fan, Q. S., Vroom, E., Récan, D. and Kaplan, J. C. (2003) 'Looking Under Every Rock: DMD and Traditional Chinese Medicine', *Neuromuscular Disorders*, 13, 705–7.

Venter, J. C., Adams, M. D., Myers, E. W., Li, P. W., Mural, R. J., et al. (2001) 'The Sequence of the Human Genome', *Science*, 291, 1304–51.

Verma, I. C. and Bijarnia, S. (2002) 'The Burden of Genetic Disorders in India and a Framework for Community Control', *Community Genetics*, 5, 192–6.

Vijay, K. (2007) 'Bangalore: Emerging Hub for Healthcare', available at: http://www.expresshealthcaremgmt.com/20020731/bangalore1.shtml (accessed 27 October 2008).

Wainwright, S., Williams, C., Michael, M., Farsides, B. and Cribb, A. (2006) 'Ethical Boundary-work in the Embryonic Stem Cell Laboratory', *Sociology of Health and Illness*, 28, 6, 732–48.

Waldby, C. and Mitchell, R. (2006) *Tissue Economies. Blood, Organs, and Cell Lines in Late Capitalism* (Durham, NC; London: Duke University Press).

Wang, J. (1995) 'International Relations Theory and the Study of Chinese Foreign Policy: A Chinese Perspective', in Robinson, T. W. and Shambaugh, D. (eds), *Chinese Foreign Policy: Theory and Practice*, pp. 481–505 (Oxford; New York: Clarendon Paperbacks).

Wang, S. G. and Zhang, S. K. (2002) 'Jibing de Wenhua Yinyue' ['The Cultural Metaphor of Disease'], *Yixue yu Zhexue* [*Medicine and Philosophy*], 9, 22–5.

Wang, X., Lu, M., Qian, J., Yang, Y., Li, S., Lu, D., et al. (2009) 'Rationales, Design and Recruitment of the Taizhou Longitudinal Study', *BMC Public Health*, 9, 233.

Weber, S. and Wilson-Kovacs, D. (2008) 'Splitting Cells: Autologous Stem Cell Practices in Germany and the UK', presented at the CBAR-Workshop Cellular Spaces, Exeter, 1 July 2008.

Wei, M. Y. (2008) 'Jing-Jin Jixueku Shouci Xiang Gongzong Kaifang' ['Beijing and Tianjin Umbilical Cord Blood Bank Opens to Public'], *Xin Jing Bao*, 27 Feb.

Wertz, D. (1999) 'State-coerced Eugenics in the Postmodern World', *The Gene Letter by GeneSage*, 1 Feb.

Wertz, D. C. and Fletcher, J. C. (1998) 'Ethical and Social Issues in Prenatal Sex Selection: A Survey of Geneticists in 37 Nations', *Social Science and Medicine*, 46, 2, 265.

Winickoff, D. (2008) 'From Benefit Sharing to Power Sharing', available at: http://escholarship.org/uc/item/845393hh (accessed 5 December 2013).

World Health Organization (1998) Proposed International Guidelines on Ethical Issues in Medical Genetics and Genetics Services, Section VI, available at: www.who.int/genomics/publications/en/ethicalguidelines1998.pdf (accessed 5 December 2013).

Wu, L. (2006) *Kexue Taijiao: Boabao gen Congming* [*Scientific Fetal Education: Even More Intelligent Babies*] (Beijing: Renmin Junyi Chubanshe).

Wu, Y. (2007) 'Discrimination Rife Against Carriers of Hepatitis Virus', *China Daily*, 23 Mar.

Wu, J. and Walther, C. S. (2006) 'Patterns of Induced Abortion', in Poston Jr., D. L., Lee, C.-F., Chang, C.-F., McKibben, S. L. and Walther, C. S. (eds) *Fertility, Family Planning, and Population Policy in China* pp. 23–37 (London: Routledge).

Xiong, L. and Wang, Y. (2001) 'Ling Rensheng yi de Guoji Jiyin Hezuo Yanjiu Xiangmu' ['Suspicious International Collaboration in Joint Genetic Research Projects'], *Liaowang* [*Outlook*], 13, 26.

Xue, J. (2005) *Chanqian Chanhou: You Wen bi da* [*Pre-birth Post-birth: Questions That Must be Answered*, transl. as *Nurture Book for Fresh Mothers*] (Changchun: Jilin Kexue Jishu Chubanshe).

Xue, Z. (2006) *Yun qi* [*Pregnancy period*] (Shijiazhuang: Hebei Kexue Jishu Chubanshe).

Yan, H. (2003) 'Neoliberal Governmentality and Neohumanism: Organizing Suzhi/Value Flow Through Labor Recruitment Networks', *Cultural Anthropology*, 18, 4, 493–523.

Yang, Y. (2003) 'Advances in Science and Progress of Humanity: A Global Perspective on DNA', in Knoppers, B. (ed.) *Population and Genetics: Legal and Socio-Ethical Perspectives*, pp. 395–404 (Boston, MA; Leiden: Martinus Nijhoff Publishers).

Yang, H.-I. (2007) 'The Use and Misuse of Informed Consent in Taiwan Biobank Debate', in *Annual Meeting of The Law and Society Association*, Berlin, Germany, 25 July 2007.

Yang, J. (2008) 'The Power Relationship Between Doctors, Patients and the Party-state Under the Impact of Red Packets in the Chinese Health-care System', unpublished PhD, University of New South Wales, Australia.

Yomiuri Online (2005) 'Shift in Aid Emphasizes Self-reliance' available at: http://www.yomiuri.co.jp/kyoiku/learning/editorial/20051021/ [in Japanese] (accessed 5 December 2013).

Yoshizumi, K. (1995) 'Marriage and Family: Past and Present', in Fujimura-Fanselow, K. and Kameda, A. (eds) *Japanese Women*, pp. 183–98 (New York: The Feminist Press).

Yuasa, J. (2007) *Shinyaku Kudasai! Duraggu Ragu to Inochi no Hazami de* [*New Medicine Please! The Abyss Between the Drug Lag and Life*] (Tokyo: Shinchosha).

Zhang, J. (2006) *Huaiyun Yangtai* [*Pregnancy and Fetal Education*] (Beijing: Renmin Ribao Dabianshe).

Zhang, Q. (2007) *Xinhun Bidu* [*Must-read for Newly Married Couples*] (Beijing: Zhongguo Renkou Chubanshe).

Zizek, S. (1991) *Mapping Ideology* (London; New York: Verso).

Index

abortion, 25–6, 27, 52
  decision-making in India, 38, 40,
    43, 44
  decision-making in Japan, 44–50
  Jôdo Shinshu view, 124
  legal status in Japan, 49, 122,
    125, 131
  thalassaemia-affected fetuses, 67–8
Agaria caste, Orissa, India, 61, 86–7
Alliance for Biosafety and Bioethics,
  South Korea, 148
All India Institute of Medical Sciences
  (AIIMS), 163
Anthropological Survey of India, 84,
  85–6
Anti-Eugenics (Thought) Network,
  Japan, 124, 151–2, 158, 206
artificial immunology (AI) technology,
  182–3
Asia
  life science hubs, 4–5
  meaning within this book, 14
assemblage, 11
Australia/Indonesia Medical Research
  Initiative (AIMRI), 99–100
autocratic decision-making, 135, 136
autonomous decision-making, 28

bad science, 128, 130, 132
Bandung Institute of Technology
  (ITB), Indonesia, 95, 96
Bangalore, Y Group of Hospitals,
  166–7
Baylis, Françoise, 136
Beck, Ulrich, 138
benefit sharing, 90–1, 104
Bhil-Pawara tribe, Maharashtra, India,
  61, 62, 64, 86–7
biobanks, human genetic, 5, 80–107
  case studies, 84–105
  regulation, 81, 197–9
  research studies using, 81–2
  uses, 80–1

biobureaucratic institutions, 54, 55,
  57, 195
biocapital, 13, 160
biocitizenship, 195
bioethical boundary-making, 111,
  134, 188–9
  concept, 109, 110, 126
  hESR in China and Japan, 126–33
bioethical governance, 7–8
bioethical problematisation, 134,
  199–200
  concept, 109–10
  reproductive cloning in China,
    111–17
  therapeutic cloning and hESR in
    China and Japan, 111, 118–26
bioethical vacuum, 7, 189–90
bioethics, 1–2
  approaches within life assemblages,
    2–3
  capacity building, 3–5, 206
  dislocation of, 203
  global moral economies and, 5–7
  institutionalisation, 137, 190,
    201–2
  life assemblages and, 188
  and medical ethics, Chinese text-
    books, 112–16, 117
  mismatch with broader life ethics,
    192–3
  'native' vs universal Western, 108–9
  pressure to apply 'global' standards,
    197–9
  professionalisation, 146–50, 201–2,
    205–6
  standardisation, 205–6
Bioethics Advisory Committee, South
  Korea, 148
Bioethics Biosafety Act 2005, South
  Korea, 149
Bioethics Expert Committee (BEC),
  Japan, 123, 124
biological weapons, 115

233

Printed and bound by CPI Group (UK) Ltd, Croydon, CR0 4YY